SYMPOSIUM

ON

HIGH TEMPERATURE COMPOSITES

SYMPOSIUM
ON
HIGH TEMPERATURE COMPOSITES

PROCEEDINGS
OF THE

*AMERICAN SOCIETY
FOR COMPOSITES*

JUNE 13–15, 1989
STOUFFER CENTER PLAZA HOTEL
DAYTON, OHIO

TECHNOMIC
PUBLISHING CO., INC.

LANCASTER · BASEL

Symposium on High Temperature Composites
a **TECHNOMIC**®publication

Published in the Western Hemisphere by
Technomic Publishing Company, Inc.
851 New Holland Avenue
Box 3535
Lancaster, Pennsylvania 17604 U.S.A.

Distributed in the Rest of the World by
Technomic Publishing AG

Printed in the United States of America
10 9 8 7 6 5 4 3 2 1

Main entry under title:
  Symposium on High Temperature Composites—Proceedings
  of the American Society for Composites

A Technomic Publishing Company book
Bibliography: p.

Library of Congress Card No.
ISBN No. 87762-700-2

Table of Contents

Preface

As our understanding of the basic issues and response characteristics of fiber-reinforced composite materials grows, a broadening of the range of usage of composites and different combinations of constituent materials are envisioned. The emphasis of the present conference and proceedings is the behavior of composites that can be designed to withstand very high temperatures, such as 1000–3000 degrees Fahrenheit, in their applications. Examples of these applications include components of turbine engines, rocket motors, and space structures. The materials receiving considerable attention for use under these challenging environmental conditions are metal matrix, ceramic matrix, and carbon-carbon composites.

Many interdisciplinary issues arise in connection with the development and usage of these modern systems, such as mechanics modeling, experimental measurement techniques, processing analysis and control, oxidation resistance, interface analysis, modification, and characterization. Because of its interdisciplinary emphasis, the American Society of Composites is well suited to provide a forum for the communication and dissemination of information on research in high temperature composite materials.

Due to the sensitive nature of some of the research in high temperature composites, a closed session was held the third day of the conference. Only presentations given at the open sessions are published in these proceedings.

The symposium chairman and officers of the society would like to express their thanks to the speakers and attendees for making the symposium on High Temperature Composites successful.

Development of an Elevated Temperature Test Technique for Metal Matrix Composite Materials

WILLIAM B. GIANNETTI

ABSTRACT

An elevated temperature test system was designed and fabricated for the determination of tensile properties of metal matrix composite (MMC) materials in inert and oxidative environments. Particular emphasis was placed on the choice of components necessary for the determination of tensile stress–strain relationships. Radiant heating was shown to produce acceptable temperature uniformity while offering the advantages of reduced specimen size requirements and low thermal mass. Factors critical to specimen geometry and gripping, strain measurement, and temperature monitoring are discussed. A microcomputer aided data analysis and reduction system is described for performing real-time monitoring and data manipulation. A material of known elastic properties at elevated temperature was tested to examine the validity of test method. Representative tensile properties were generated for SCS–6/Ti 15–3 MMC at elevated temperature in air and argon environments to assure total system functionality.

INTRODUCTION

Metal matrix composites (MMC's) are unique because they offer the high strength, stiffness, and environmental stability of the reinforcement fiber, and also retain the elastic-plastic behavior of the monolithic matrix material. They offer a higher temperature capability then organic matrix composites and better fracture and fatigue behavior then ceramic matrix composites. Because of these characteristics, MMC's have received a great deal of attention as candidate materials for such programs as the National Aerospace Plane (NASP) [1], and Integrated High Performance Turbine Engine Technology (IHPTET) [2]. In these challenging applications structural materials are expected to maintain strength and dimensional stability at temperatures exceeding 1500°F, thus eliminating all but a few monolithic metals and alloys as practicable candidates for weight sensitive primary structural elements. Of these few monolithic materials (e.g. titanium and titanium aluminide alloys), none have sufficient strength and stiffness at elevated temperatures to maintain the structural requirements set forth by these programs. However, selective reinforcement of these alloys with high strength, high stiffness ceramic or graphite fiber results in a MMC with the capability to meet the basic elevated

W.B. Giannetti, Materials Engineer, Textron Specialty Materials, 2 Industrial Avenue, Lowell, Massachusetts 01851 temperature design criteria.

Each type of engineering material has a unique set of behavioral characteristics and MMC's are no exception. At elevated temperature MMC's exhibit a full spectrum of deformation behavior ranging from linear-viscoelastic to elastic-perfectly-plastic depending upon the particular fiber loading and thermal environment. During fabrication the constituent materials interact to produce both thermal residual stresses [3] and interfacial reaction zones [4,5]. The thermal residual stresses result from coefficient of thermal expansion mismatches between the fiber and the matrix causing residual compression in the fiber and residual tension in the matrix. The interfacial zone can exist in either tension or compression depending upon its elastic properties. It is preferable to have the matrix material in residual compression because yielding of the matrix is a limiting factor in MMC's at elevated temperature. The interfacial zone forms as a result of the chemical interaction between the constituents and is responsible for effective load transfer between the matrix and the fiber.

Several methods are available for analyzing the behavior of MMC's at elevated temperature. Micromechanics [6] and micro-macromechanics [7] models have been developed which account for mechanical property predictions, thermal properties, and constituent microstresses. An alternative method is to perform elevated temperature destructive testing. This method serves two purposes, one being to confirm theoretical mechanics predictions, the other to evaluate the behavioral characteristics of a material system under a variety of simulated service conditions. These conditions often exist outside the realm of theoretical analysis where for example microstructural behavior is dependent upon environmental conditions (i.e. oxidation degradation, matrix phase transformation, etc.)

In order to fully understand the behavior of materials various thermal and environmental conditions, it is imperative that accurate and versatile elevated temperature test systems be established. Unfortunately, little has been done in the field of elevated temperature composite material test standardization and only recently has an ASTM committee (C-28) been formed to address these issues with respect to ceramics matrix composites. This lack of standardization can be attributed to several factors, the primary one being that test methods are often tailored to a specific material system and test requirements leading to a myriad of custom test system design approaches.

APPROACH

The design approach for the elevated temperature tensile test system (ETTTS) described in this paper was dictated by attempting to satisfy the following criteria:
(1) Maximize utility of test system to perform under a variety of test conditions.
(2) Minimization of sample material cost
(3) Rapid turnaround of test results

Each criterion dictates the diligent choice of the six specific components of the overall system which need to be considered in ETTTS design: the load frame, heating system, temperature monitoring system, strain sensor, gripping system, and environmental chamber. Often the choice of one component dictates the choice of the other making the entire design process iterative as indicated by Gyekenyesi and Hemman [8] in designing an ETTTS for ceramics composites testing.

LOAD FRAME For this design a United DM-60 twin screw universal tensile test system was used. The load frame has an upper limit load capability of 60,000 pounds and a manual crosshead rate controller with digital readout. The loadcell used has a linear voltage output of 0-30 volts full scale in tension. This load frame does not allow for fatigue mode testing, but with slight fixturing modifications, the ETTTS could easily be retrofit to a servohydraulic machine for a fatigue/thermofatigue testing.

HEATING SYSTEM The major requirement of the heating system is to assure a uniform temperature over a fixed length of the specimen being tested. This fixed length is necessary for accurate strain measurement and is dictated by the gage length of the extensometer. The estimated maximum operating temperatures for metal matrix composites range from 600°F to 1800°F. It is therefore essential that materials be characterized up to these temperatures. Several heating methods are available which can easily manage this range of temperatures [9], but that availability disappears when additional requirements are imposed for MMC testing: the specimens be subjected to inert and oxidative environments, the heating rate be suffi-ciently high (<30°F/sec) and therefore the thermal mass of the furnace be sufficiently low to accommodate controlled thermal cycling and thermal spike experiments, the tensile specimens be 8 inches in length, and the entire system, furnace and environmental chamber, be capable of fitting within the available load frame.

One method that has been shown to satisfy these requirements is radiant heating [10] wherein infrared energy can be directed to highly localized regions of the sample. This method of heating and the choice of offset heaters, which will be described below, satisfies all three of the objective design criteria.

The furnace system for this design consists of two model 5066-3 radiant patch heaters manufactured by Research Inc. Each radiant heater contains three type Q500T4/2CL Sylvania quartz lamps. The lamp filament operates at approximately 4500°F at rated voltage with an 1.0 micron spectral energy peak. Each lamp outputs an average of 500 watts giving a total rated output of 1500 watts per heater. The total heat flux at rated voltage is estimated to be 565 watts/sqin.

To prevent the quartz tubes and aluminum manifold from melting, forced air and water are introduced into the lamp housing at 30 psig and 0.8 Gpm, respectively, to create a forced convection cooling environment.

The radiant heaters are mounted vertically opposing each other to maximize the temperature capability while minimizing the thermal gradients through the thickness of the specimen. The mean separation between heaters is 0.2 inches to allow for non-obstructed specimen and extensometer inser-tion and extraction. In this configuration several conditions are simulat-ed, for example, aerodynamic heating effects where only one side of a specimen, say a potential wing skin MMC, is heated.

Experiments were conducted to determine the feasibility of obtaining a uniform temperature distribution, to within +/- 25°F, over an effec-tive length of one inch. This one inch specification is necessary for accurate strain measurement and is governed by the separation distance between contact points of the extensometer (this will be referred to as the effective gage length (EGL)).

Initial results with the as supplied heaters were inadequate for the accurate determination of strain due to a 400°F gradient, at a peak specimen temperature of 1800°F, within the EGL of the specimen. An acceptable temperature distribution was obtained by modifying the incident radiant flux on the specimen by (1) rewiring the heaters to allow for inner and outer zone control and (2) placing silicon carbide dividers between the individual lamps to minimize radiant crosstalk and undesirable radiant reflection onto the specimen. Fig.1 shows the temperature distributions within the EGL of a 304 stainless steel specimen. As anticipated, the temperature distribution is highly uniform below 1000°F and takes on a characteristic parabolic shape (due to net heat transfer by conduction to the grips) above 1000°F. As the temperature approaches 1800°F the heat transfer to the grips becomes more apparent leading to a slight hot spot in the center of the specimen. Above 1800°F the temperature uniformity exceeds the allowable tolerance.

TEMPERATURE MONITORING SYSTEM The challenge of accurately determining temperature of the specimen during a test was investigated from several different approaches. Thermocouples imbedded within the specimen gave accurate temperature distribution results but could not be used during actual testing. This method did serve, however, to check results of different techniques described below. Infrared pyrometers were investigated but proved deficient with respect to optical requirements or applicable sensor frequency ranges. Color match pyrometers were considered, but since they are manually operated, they could not be considered in a closed loop temperature control scheme. Finally, the bonding of thermocouples to the specimen proved feasible.

Experiments with various bonding agents (organic, carbon, silver) were conducted and temperature profiles determined. These materials served not only to bond the thermocouple to the material, but also acted as a coating to minimize errors due to radiant cooling [11]. Results when compared to the control thermocouple indicated that the carbon coated bead followed the control thermocouple within 15°F to 1800°F. Bonding problems were encountered, however, at the higher temperatures. Thermocouples welded to the specimen and then coated with the carbon binder proved the best solution. Stress concentrations in the immediate vicinity of the weld were minimum due to the small size of the thermocouple bead (0.005 inches in diameter) based on the random location of specimen failures.

STRAIN SENSOR Most contact extensometers are configured with two horizontal rods separated by a fixed vertical distance (gage length). The one major disadvantage of this type of extensometer is that the force used to counteract the weight of the rods produces moments in the same plane as the mounting force used to press the rods against the specimen [8] . One contact extensometer designed with precision balanced rods, thus eliminating the undesirable moments, was chosen for this application.

The strain sensing system chosen for this application was an Instron 1500°C capacitive extensometer. It is an integral part of the overall test system but is mounted outside the environmental chamber to minimize thermal effects which may alter instrument precision. The extensometer rods are loaded perpendicular to the edge of the specimen to accurately track the vertical deflection of the material. Specimen tracking is assured by a precision balancing of the contact rods to eliminate the resistance to vertical movement. The contact force for this instrument is produced by a three gram spring loading perpendicular to the specimen surface. All positioning devices and deflection monitoring electronics are external to the environmental enclosure for accessibility purposes. A linear voltage is generated by the extensometer which is proportional to the separation distance between two capacitive plates mounted at the end of the contact rods. The extensometer has a linear range of 0.04 inches in tension and compression.

GRIPPING SYSTEM   The primary goal of the gripping system is to apply uniform forces to the material sample with the objective of uniform stresses in the test section. Screw type wedge gripping systems often apply direct compression to the specimen which resolves into shear and compression components at the specimen/grip interface. The major disadvantage of this type of grip system is that the compression component is a function of applied load and premature failures have been observed in high strength MMC due to specimen crushing at the grip interface [12]. Gripping systems can also cause stress concentrations on the material sample due to bending or irregularities at the grip to specimen interface. To avoid this, grip to grip alignment as well as grip and specimen surface uniformity are critical for achieving accurate results.

Two series 647-10 Hydraulic Wedge Grips manufactured by MTS Systems Corporation were chosen for this application. The grips have been specifically designed for static and/or fatigue mode testing application thereupon satisfying the first design criterion. The grips provide a constant, hydraulically actuated gripping force regardless of the applied test loads. The specimen gripping force is adjustable to prevent either specimen crushing or slippage during a test. The grips have an upper load capacity of 20,000 pounds.

Cold gripping (i.e. gripping outside the hot zone) was chosen to eliminate many of the problems associated with hot gripping, i.e. chemical reaction between the grips and the specimen, grip oxidation, and grip induced failures. With cold grip systems, failures are localized within the hot zone where the material is in a state of thermal strain.

The hydraulic fluid used is a high temperature grade Mobil 525 SHC with an upper temperature limit of 325°F. This limit specifies the maximum operating temperature of the grips. Experiments indicate that the maximum wedge face temperature during a 2000°F test reached 275°F, following a 30 minute specimen soak, while the grip body temperature increased by only 25°F.

The grip faces are made of a material known as Surfalloy. This material offers excellent gripping with minimal pitting to the specimen surface.

ENVIRONMENTAL ENCLOSURE   Titanium alloys have limited oxidation resistance in air at temperatures above 1000°F [13]. Recent research with a titanium 15-3 composite indicated that the formation of a thin oxide film of titanium dioxide may contribute to a large decrease in tensile strength and ductility [14].

To minimize the effects of oxidation and also to satisfy the first design criterion, an environmental enclosure was designed to encapsulate the grips and heaters. The enclosure design allows for 0.01 Torr vacuum and inert gas backfill (UHP argon). The repeated cycling of vacuum/argon followed by a positive pressure backfill of argon during an elevated temperature test has been shown to minimize the available oxygen in the test system and therefore oxidation.

DATA ANALYSIS AND REDUCTION SYSTEM   A microcomputer aided data acquisition system was configured and integrated into the overall test system. With this capability, scaled parameters such as temperature, time, stress, strain, and modulus can be monitored in real-time during an actual elevated temperature test.

A photograph and schematic diagram of the overall test system are shown in Figures 2(a) and 2(b), respectively. A closeup of the test system showing the various components is presented in Figure 3. The 16 bit microcomputer is equipped with an Intel 8087 math coprocessor, 640 Kb of RAM memory, 20 Mb hard disk, and an EGA color monitor. An HP 3497A data acquisition and control system is connected to the computer via IEEE-488 GPIB (General Purpose Interface Bus) parallel interface card. All data acquisition and reduction routines were configured using Labtech Notebook, a multipurpose data acquisition and control software program.

During a typical test, voltages from the loadcell, strain sensor, and thermocouples are monitored at a predefined sampling frequency. For this application 2 Hz was sufficient. The software assigns each voltage to a channel which can then be addressed by the software. In all, there are eight input voltage channels, six for thermocouples, one for load, and one for strain. The voltages are converted to meaningful data by the use of scale parameters and offset constants which are determined by the characteristics of the equipment.

In data processing, it is necessary to obtain not only the maxima, i.e. ultimate strength, ultimate strain, but also the derivative of the test data for modulus determination. The procedure used here involved a smoothing operation where the original stress and strain data are smoothed using a three point averaging technique. The smoothed stress and strain are then differentiated directly by dividing into equal segments and fitting a straight line in each segment. This technique eliminates the noise associated with the point-by-point technique i.e. instantaneous tangent modulus. Figure 4 depicts a simplified flowchart for the tensile test data processing program, where all operations, including data acquisition, smoothing, differentiating, data storage, plotting are performed simultaneously. In response to the operators need, the processed data can displayed on the monitor screen in the form of digital meters, stress-strain curves, etc..

EXPERIMENTAL

Initial system trials were carried out at room temperature in order to evaluate the mechanics of the test system. Both aluminum and steel standards were tested to determine the accuracy and sensitivity of the system with respect to strain and load tracking. Initial trials showed good agreement with textbook values when comparing Young's modulus.

Qualification of the ETTTS was accomplished by testing a material of known elastic properties at elevated temperature. Young's modulus was chosen as a performance measure for two reasons, (1) it is an intrinsic material property, and (2) it provides good verification of strain measurement capability. Annealed austenitic 304 stainless steel was selected as a reference material due to the availability of published mechanical property data at elevated temperature [15]. In this test series, 15 specimens were machined from plate stock to form tensile coupons. The specimens were straight-sided, constant cross section, having the following nominal dimensions: 0.087 x 0.500 x 8.000 inches, indicating small material utilization. Three specimens were tested in air at each of five temperatures; RT, 800, 1000, 1200, and 1500°F. Specimens were held at temperature for five minutes to assure thermal equilibrium. A comparison of the results determined experimentally against those published, as shown in Table 1, indicates that the data sets are well within the bounds of experimental scatter.

Three panels of four ply unidirectionally reinforced SCS-6/Ti (15-3) were investigated for short term elevated temperature tensile properties following oxidative and inert environmental exposures. Of the three panels, two were fabricated to have volume fractions of approximately 55%, the third 35%. All three panels experienced the same processing history thus minimizing the effect of process variability on the test results.

Prior to specimen machining, the panels were subjected to an extensive nondestructive evaluation to determine the location of internal anomalies. The evaluation consisted of water immersion C type scans as well as radiographic examinations. Following specimen machining, the samples were re-examined radiographically using Tetrabromoethane (TBE), a radio-opaque penetrating fluid, to determine both the extent of machining damage and the location of internal flaws not observed in the initial investigation.

The test specimens were machined from 8x8 inch panels using a diamond wheel cutting technique. All specimens were cut with the fibers aligned in the axial direction. The specimen geometry used for this evaluation were straight sided with nominal dimensions (in inches) and estimated fiber volume percent given in Table 2.

Prior to testing, the specimen dimensions were measured and the values recorded for subsequent input into the data reduction program. Type S thermocouples were attached to the specimen for temperature monitoring. The specimens were manually inserted and the lower grip closed to assure placement. The upper grip was left open to allow for stress free thermal expansion during the heat-up process. Once the desired temperature was stabilized, the upper grip was closed and the extensometer introduced. Following a five minute exposure at temperature, the cross head was started and data acquisition commenced. Argon exposure experiments were conducted in a similar manner as the air exposed but with an initial evacuation of the test chamber followed by a positive pressure argon purge.

Young's modulus results were obtained for the 55% volume fraction SCS-6/Ti (15-3) MMC at temperatures up to 1800°F following a five minute exposure at temperature in air. Results were also obtained for the 35% volume fraction SCS-6/Ti up to 1500°F following a five minute exposure in air and argon environments. The results of these tests are presented in Tables 3(a) and 3(b) and in Figure 5. Elevated temperature Young's modulus data for SCS-6 fiber exposed to argon [16] are included in Table 3(b) to demonstrate the similar modulus reduction with increasing temperature for both fiber and composite.

Elevated temperature Young's modulus determined by Watson [14] on a 43 volume percent SCS-6/Ti 15-3 MMC following a half hour soak in air at temperature is included in Fig 5 for comparative purposes. One possible explanation for the discrepency in modulus at the higher temperatures is the demonstrated microstructural sensitivity of the modulus of the matrix material [17]. The modulus in Ti-15-3 has been reported to vary between 11 and 14 Msi depending on the thermal-mechanical history of the alloy.

CONCLUSIONS

Proof of system design and repeatability were demonstrated on annealed austenitic 304 stainless steel with a comparison of experimental to published values.

System viability for determination of tensile properties of MMC's was demonstrated on silicon carbide reinforced titanium (SCS-6/Ti 15-3) in oxidative and inert environments at elevated temperature.

Based on the judicious choice of components, the following capabilities have been demonstrated:

(1) Load- strain tracking capability over a wide range of temperatures with flexibility over choice of environment.
(2) Small material utilization.
(3) Good temperature control.
(4) Potential for thermal cycling, thermal spike and fatigue testing.

Since this project was centered around the development of an elevated temperature tensile test system, no conclusions will be drawn concerning the degradation mechanisms observed in the SCS-6/Ti (15-3) material. The testing was done for the sole purpose of generating data on a representative metal matrix composite material system.

REFERENCES

1.   Tortolano, F.W.,"Birth of a Space-Age Plane," Design News, Apr 1988, pp. 52-58.

2.   U.S. Congress, Office of Technology Assessment, New Structural Materials Technologies: Opportunities for the Use of Advanced Ceramics and Composites-A Technical Memorandum, OTA-TM-E-32, Washington,D.C.; US Govt Printing Office, Sept 1986.

3.   Pederson, O.B.,"Residual Stresses and the Strength of Metal Matrix Composites," Metallurgy Dept., Riso National Laboratory, Denmark, 1986.

4.   Jones, C., Kiely, C.J., Wang, S.S.,"The Characterization of an SCS-6 Ti-6Al-4V MMC Interface," Journal of Materials Research, Vol.4, No.2, Mar/Apr 1989, pp. 327-335.

5.   Lerch, B.A., Hull, D.R., and Leonhardt, T.A.,"As-Received Microstructure of a SiC/Ti 15-3 Composite," NASA TM-100938, Aug 1988.

6.   Hopkins, D.A., Chamis, C.C.," A Unique Set of Micromechanics Equations for High Temperature Metal Matrix Composites," NASA TM-87154, Nov 1985.

7.   Bahei-El-Din, Y.A.,"Plasticity Analysis of Fibrous Composite Laminates Under Thermomechanical Loads," Proc of the ASTM Symp on Thermal and Mechanical Behavior of Ceramic and Metal Matrix Composites, Nov 1988.

8.   Gykenyesi, J.Z., Hemann, J.H.,"High Temperature Tensile Testing of Ceramic Composites," Cleveland State University, Cleveland, Ohio, Jun 1987.

9.   Halford, G.R.,"Low cycle Thermal Fatigue," NASA TM-87225, NRLC, Cleveland, Ohio, Feb 1986.

10. Hartman, G.A., and Russ, S.M.,"Techniques for Mechanical and Thermal Testing of Ti3Al/SCS-6 Metal Matrix Composite," Metal Matrix Composites: Testing, Analysis, and Failure Modes, ASTM STP 1032, W.S. Johnson, Editor, ASTM, Philidelphia, 1989.

11. Holman, J.P., Heat Transfer, Fourth Edition, McGraw Hill Book Company, New York, 1976, p. 341.

12. Giannetti, W.B.," Tensile Testing of Large Cross Section Silicon Carbide Reinforced Aluminum," Textron Internal Report, Apr 1987.

13. MIL-HDBK-5B, Aug 1973.

14. Watson, D.C.,"Effect of Elevated Temperature Exposure on the Tensile Properties of SCS-6 Titanium Composite," AFWAL-TR-88-4170, Materials Laboratory, Wright Patterson Air Force Base, Ohio, Oct 1988.

15. Mechanical and Physical Properties of the Austenitic Chromium-Nickel Stainless Steels, Published by the International Nickel Company, Inc., 3rd Edition, Nov 1968.

16. Giannetti, W.B.,"Development of an Elevated Temperature Test Technique for Composites," Textron Specialty Materials, IRAD Proj. No 87009492, 1987.

17. TIMET Ti-15-3 Data Sheet, Pittsburgh, PA, 1982

TABLE 1. Comparison of elevated temperature Young's modulus (in Msi) for annealed austenitic 304 stainless steel.

TEMPERATURE (F)	EXP DETERMINED MODULUS	PUBLISHED MODULUS
RT	28.92	28.3
800	24.17	24.1
1000	22.22	22.5
1200	20.86	21.1
1500	18.44	18.1

TABLE 2. Nominal dimensions in inches for SCS-6/Ti (15-3) specimens.

Volume Fraction	Width	Thickness
.35	.25	.053
.55	.25	.035

TABLE 3. Elevated temperatue Young's modulus of SCS-6/Ti (15-3) specimens exposed to (a) air environment, (b) argon environment, with inclusion of SCS-6 fiber.

(a)

Test Temp (F)	35% Vf Tensile Modulus (msi)	55% Vf Tensile Modulus (msi)
72	27.4	35.6
800	26.0	31.8
1000	25.2	30.6
1200	24.4	30.5
1500	22.1	28.2
1800	---	27.6

(b)

Test Temp (F)	35% Vf Tensile Modulus (msi)	SCS-6 Fiber* Tensile Modulus (msi)
72	27.4	58.0
800	30.3	55.6
1000	28.9	54.9
1200	28.0	54.2
1500	26.1	53.2
1800	---	52.2

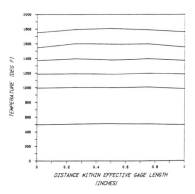

FIGURE 1. Temperature distribution within EGL of 304 SS specimen showing temperature uniformity up to 1800°F

(a)

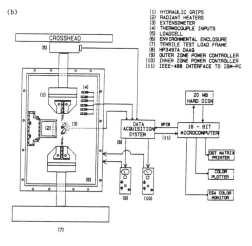

(b)

(1) HYDRAULIC GRIPS
(2) RADIANT HEATERS
(3) EXTENSOMETER
(4) THERMOCOUPLE INPUTS
(5) LOADCELL
(6) ENVIRONMENTAL ENCLOSURE
(7) TENSILE TEST LOAD FRAME
(8) HP3497A DAAQ
(9) OUTER ZONE POWER CONTROLLER
(10) INNER ZONE POWER CONTROLLER
(11) IEEE-488 INTERFACE TO IBM-PC

FIGURE 2. (a) Photograph, (b) schematic diagram of elevated temperature tensile test system.

FIGURE 3. Close up photograph of critical tensile test system components.

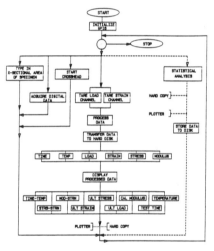

FIGURE 4. Simplified flowchart of tensile data processing system.

YOUNG'S MODULUS (MSI)

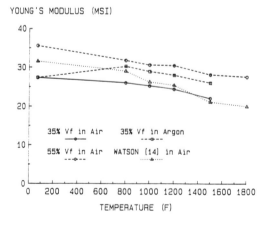

FIGURE 5.   Axial Young's modulus behavior of SCS-6/Ti (15-3) MMC at elevated temperature
in air and argon environments with comparison to Watson [14].

10

Interfacial Modification in G/Al Metal Matrix Composites

J. P. CLEMENT AND H. J. RACK

ABSTRACT

Successful development of metal matrix composites has been limited by two recurrent problems: reinforcements are difficult to wet by molten metals, and chemical reactions take place at the fiber-matrix interface.

Detailed examination of thermodynamic and kinetic data suggest that it should be possible to develop either a matrix composition or interfacial barrier coating which promotes wetting, while simultaneously reducing interfacial interaction.

The current study has developed a procedure to coat carbon reinforcements with a thin, crack-free, protective oxide layer using a sol-gel technique, emphasis in this investigation being placed on TiO_2 coating for inclusion in an aluminum metal matrix.

The effect of the solution composition and coating procedure on the coating uniformity, structure, and thickness were investigated. Fiber coating were performed utilizing high strength PAN base carbon fibers, with the coated fiber preforms being infiltrated by pure liquid aluminum. Subsequent examination included energy dispersive x-ray analysis, optical and transmission electron microscopy of the fiber-matrix interface.

INTRODUCTION

Continuous carbon fiber-aluminum alloys, if properly integrated, form a class of metal matrix composite materials whose properties may be tailored to satisfy many demanding structural requirements. These light-weight composite materials exibit high strength and stiffness, high thermal and electrical conductivities, and do not outgas in a vacuum. Unfortunately, carbon is difficult to wet with liquid aluminum at moderate temperatures, that is at less than 500°C, while at higher temperatures Al_4C_3 formation occurs at the fiber-matrix interface [1]. Formation of the latter causes the strength of the composite to decrease [1].

TiB_2 coating applied by Chemical Vapor Deposition (CVD) is the approach currently used to prepare continuous C-Al composites [1,2]. Although CVD processing has been sucessful it is difficult to obtain a uniform coating around the circumference of each fiber using this process.

J.P. Clement, formerly Graduate Student, now Materials Research Engineer, Aerospatiale TX, 78130 Les Mureaux, France, and H.J. Rack, Professor of Mechanical Engineering and Metallurgy, Clemson University, Clemson, SC 29634-0921.

The fibers must be well separated in order to avoid the problem of shadowing--when one fiber overlaps another and prevents it from being coated properly. In addition, this coating process is carried out at high temperature, and usually degrades the carbon fibers.

Another serious problem with a TiB_2 coating is that it is not air stable. The coated fibers cannot be exposed to air before immersion in the molten aluminum, otherwise wetting will not take place [1,2].

Katzman [2] has recently suggested an alternative approach for the preparation of C-Al composites, one utilizing SiO_2 sol-gel fiber coatings. This investigation has extended this effort focussing on TiO_2 sol coatings. TiO_2 was selected based upon its thermodynamic stability with respect to TiC formation, the known ability of Ti to enhance the wettability between metals and ceramics, and its ready availability in the form of a metal alkoxide precursor, titanium isopropoxide (TIP) [3,4].

The objectives of this research were to establish parameters for producing thin uniform TiO_2 sol coatings on carbon fibers, to demonstrate the ability of aluminum to wet the TiO_2 surface, to evaluate the effectiveness of the coating as a diffusion barrier to prevent Al_4C_3 formation, and finally, to examine any interfacial reaction products.

EXPERIMENTAL PROCEDURE

Sol preparation involved combining selected amounts of water and hydrochloric acid in a beaker containing 2-propanol and thoroughly mixing for 5 min to obtain the desired concentration. The required amount of titaniun (IV) isopropoxide (TIP) was then added to the solution, and the solution was stirred in a covered beaker for at least 1 hr.

Glass microscope slides, carbon coated slides, and sapphire crystals were utilized to establish the initial sol-gel coating procedure. The slides and/or unsized PAN 650/42 carbon fiber tows were dipped into the coating solution for 1 min and withdrawn vertically at a constant speed. They were then dried at 60°C and fired in air at temperatures varying from 300°C through 700°C. Some carbon fiber samples were also fired in a carbon monoxide (CO) atmosphere at 700°C.

Potential C-Al interactions were examined utilizing squeeze cast pure aluminum composites containing 12 v/o carbon fibers with the temperature of the preform being selected to minimize the infiltration pressure [5]. The cooling rate of the infiltrated samples was approximatively 7°C/min, pressures as low as 50 psi being used to infiltrate the fiber preforms.

Optical metallography was performed on samples after infiltration in order to establish the fiber dispersion and to qualitatively define the amount of reaction at the fiber-matrix interface. Composite thin foils were then prepared by mechanical thinning and ion milling, and observed utilizing a TEM JEOL 100C, and a Philips 420 equipped with a PGT EDX analysis system.

RESULTS AND DISCUSSION

Coating Evaluation

Thermal Effect

Once a substrate is coated, the heat treatment--drying and firing--is most critical. The DSC results shown in Fig. 1. revealed that drying--evaporation of the solvents--occurred between room temperature and 200°C, with maximum desorption of solvents at 125°C. Above 200°C, the desorption of the solvents was complete, the weak exothermic peak found at approximately 300°C being probably due to carbonization of the remaining OR groups [6]. A second weak exothermic peak due to the amorphous-anatase phase transformation, was also observed at approximately 420°C. The accompanying TGA observations, confirm that the maximum weight loss, 25%, occurs during drying, with the weight loss between 200°C and 400°C being smaller-- approximately 5%. Above 400°C the weight loss is negligible.

Coating Thickness

Coating thicknesses, as determined initially on glass slides, depend upon the TIP concentration of the solution, the speed of withdrawal of the samples from the sol, the number of dips of the samples in the solution, and the firing temperature, Fig. 2. For example, as the speed of withdrawal, s, increased--at least within the range of withdrawal speed examined in this investigation--the thickness, t, of the coating increased, i.e., t is proportional to s^n where n varied from 0.5 to 0.6. It was also observed that, at constant H_2O and HCl molarities, n increased as the TIP concentration increased, increasing from approximately 0.50 when the TIP concentration was 0.1 mol/l, to 0.61 when the TIP concentration was 0.5 mol/l. Finally, increasing the number of dips or the TIP concentration lead to a linear increase in the coating thickness.

Fig. 2. further suggests that the original surface energy and chemistry has an effect--albeit a seemingly secondary one--on the coating thickness. The initial dip resulted in a surface coating thickness of 25 nm, while each subsequent dip resulted in an increased thickness of 45 nm. The same phenomenon was also observed on carbon coated slides, where under identical conditions, fired sol coatings were thinner than those on clean glass slide.

Successful experiments were conducted on glass slides which produced, in a single dip, crack free TiO_2 layer as thick as 200 nm. Above 200 nm, the coating cracked during the firing step at 300°C. Above 270 nm, layers even cracked during drying. Sherer [7] suggested that during drying, stresses arise from differential strains. These stresses result from the fact that, due to differences in permeability, the exterior and the interior surfaces do not shrink at the same rate. The stresses can be quite large for organometallic gels, exceeding the coating strength. However, thin films do not crack during drying; presumably because the stresses associated with differential shrinkage are relieved by strain relaxation [7].

Coating Structure

Experiments conducted on scratched glass slides revealed that polymerization was disturbed by substrate topography. Surface roughness caused imperfections in the coating uniformity, with enhanced hydroxide particle nucleation being associated with surface imperfections.

The effects of firing for 15 min on the coating crystal structure were determined for temperatures up to 700°C. Coatings--150 nm--were prepared on sapphire substrates utilizing the procedure described earlier. Coatings were amorphous as formed, and continued to be so up to 400°C, Fig. 3. At and above 400°C, crystallinity developed, anatase being the first ordered phase observed. Mukherjee [8] suggested that the anatase phase nucleates first because it is structurally closer to the reactant, and that the rutile phase forms later. The data do not indicate the anatase-rutile phase transformation, though there appears to be some evidence of the (110) rutile peak in the 700°C data.

Carbon Fiber Coating

Carbon fibers were coated utilizing the dipping procedures previously described. The fibers were dipped as tows, 12000 filaments.

A multiple dip procedure at a lower TIP concentration was found to be much more efficient than a single dip procedure for obtaining a thin, uniform coating on the rough fiber surface. The first dip tended to smooth surface imperfections, and change surface chemistry allowing the second dip to deposit a more uniform layer. Multiple dipping also had the advantage of permitting gradual solvents evaporation during drying and firing, thereby resulting in lower residual stresses coating [7].

Because the inclusion of carbon fibers in aluminum was anticipated, the final firing temperature selected for coated fibers was 700°C. If coatings were directly heated in air to 700°C they cracked without regard to their thicknesses. When thin coatings were first fired at 400°C, and then at 700°C in air the appearance of the surface improved. This was thought due to structure homogenization throughout the coating thickness [7]. When fired directly at 700°C, at a rapid heating rate, the coating surface was presumably crystalline and the interior of the coating was still amorphous, and not as dense. Consequently, a high stress gradient was created throughout the coating thickness causing coating fracture. When fired first at 400°C, the coating structure homogenized under a smaller stress gradient, and was then better able to withstand the second firing at 700°C.

Selective firing between dips also improved the coating uniformity. Optimally the intermediate firing temperature should be kept as low as possible, i.e., 400°C, to minimize the exposure time of the coated fibers to high oxidizing temperatures. Coatings can still be fired if necessary at 700°C after the final dip.

Coatings on PAN fibers were uniform and crack free when fired in air for 15 min at temperatures below 650°C. At higher temperatures, however, coatings fractured, Fig. 4. In contrast, when fired at 700°C in a CO atmosphere the coatings did not crack. Summarizing, the best coating results were obtained using a multiple dip procedure with 400°C as the intermediate firing temperature, and 700°C in CO being the firing treatment after the final dip, Fig. 4.

The importance of firing atmosphere can be considered on a thermo-dynamic basis. Thermodynamic calculations show that in air the following reaction can take place at temperatures as low as 650°C [9].

$$TiO_2 + 3C \longrightarrow TiC + 2CO$$ In contrast, in a CO atmosphere TiO_2 is stable with respect to TiC formation at temperatures higher than 1200°C.

Stronger gels may also be formed when fired in a CO atmosphere because, in the presence of oxygen, residual terminal bonds are not forced to share oxygen and, will therefore, create a looser oxide network [10]. Thus under a CO atmosphere, the following reaction takes place, and stronger coatings were produced.

$$Ti-OH + OH-Ti \longrightarrow Ti-O-Ti + H_2O$$

These coatings are then able to better withstand the thermal stresses and the structure transformation occuring during firing.

Finally, the surface of all crack free coatings produced on fibers appeared very smooth. The coating grains were not visible under SEM, in contrast to coatings applied by sol-gel on Al_2O_3-Saffil fibers by An and Luhman [11].

Examination of the Interface Fiber-Matrix

Transverse cross sections of cast composites revealed that TiO_2 coatings enhanced the wettability of the fibers by molten aluminum, Fig. 5. In the case of non-coated fibers, the fibers were <u>not</u> dispersed in the matrix. When coated carbon fibers were used, however, the fibers were dispersed. The fiber distribution was however non-uniform due to the effect of the aluminum flow during infiltration.

Initial casting experiments were conducted with coated fibers directly fired at 700°C in air. The coating was already cracked in some places before infiltration, and a large Al_4C_3 layer, 250 nm, was formed at the interface fiber-matrix. No Ti was found close to the interface; however, some was found in the matrix away from the fibers. The precracked coating was probably separated from the fiber during the infiltration process, the fibers were then unprotected and in contact with molten aluminium causing the formation of carbides.

Another set of casting experiments was conducted using fibers coated with a 330 nm crack free film fired in CO. This coating was made of three 110 nm layers. Each dip did not cover the whole fiber surface and in fact the actual average coating thickness was close to 220 nm. Analytical TEM work and TEM diffraction patterns confirmed that the coating was a very fine grain--15 nm average--polycrystalline anatase layer, Fig. 6. A small part of the coating reacted with aluminum to form a 70 nm Al_2O_3-Ti layer, allowing chemical wetting between the coated fibers and the matrix. Notice that even a thinner coating--110 nm--prevented the carbide formation as well.

Thermodynamically the coating should have been completely reduced by molten aluminum [9].

$$3/2 \ TiO_2 + 2 \ Al \longrightarrow Al_2O_3 + 3/2 \ Ti$$

However, kinetic and diffusional effects take presidence and only a small amount of the coating was reduced, the 70 nm reaction layer preventing more species from diffusing. These findings are in agreement with work done by Katzman [2] and Chin and Numes [12] on SiO_2 sol-gel coated carbon fibers and magnesium.

CONCLUSIONS

1. A technique to coat PAN fibers with a crack free, air stable, protective TiO_2 layer by the sol-gel process has been developed.
2. Coating thickness can be controlled by the TIP concentration and the water content of the sol, the speed of substrate withdrawal, and the number of dips. Coating thickness is also influenced by the substrate geometry, chemistry, and structure.
3. Heat treatments reduce the coating thickness and change the coating structure. After heat treatment at 400°C for 15 min coatings reach their quasi-final thickness and are still amorphous. At temperatures between 400°C and 700°C the coating is transformed into a fine grain polycrystalline anatase layer.
4. 100 nm thick TiO_2 coatings enhance the wettability of PAN carbon fibers by pure molten aluminum and prevent carbide formation at the interface fiber-matrix even after 9 min in contact with pure molten aluminum.
5. Part of the TiO_2 coating reacts with aluminum to form an Al_2O_3-Ti layer allowing chemical wetting and acting as a diffusion barrier.

ACKNOWLEDGEMENT

The support of this work by Aerospatiale (JPC) and the U.S.Air Force Office
of Scientific Research, contract F49620-87-C-0017, as part of the
University Research Initiative program on High Temperature Metal Matrix
Composites at Carnegie Mellon University (HJR), is gratefully acknowledged.

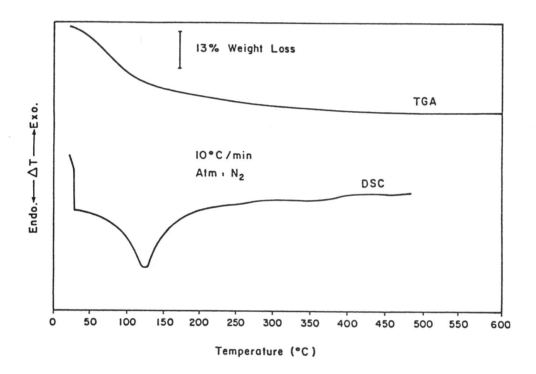

Fig. 1. Thermal analysis of TiO_2 gel dried in air at room temperature.

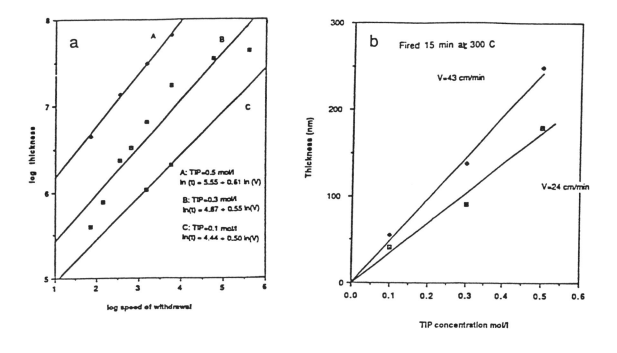

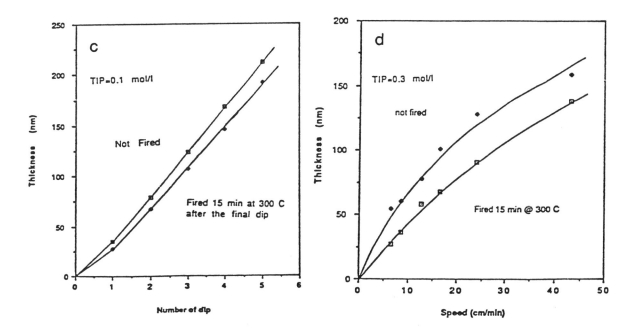

Fig. 2. Influence of (a) substrate withdrawal speed, (b) TIP molarity, (c) number of dips, and (d) firing on coating thickness.

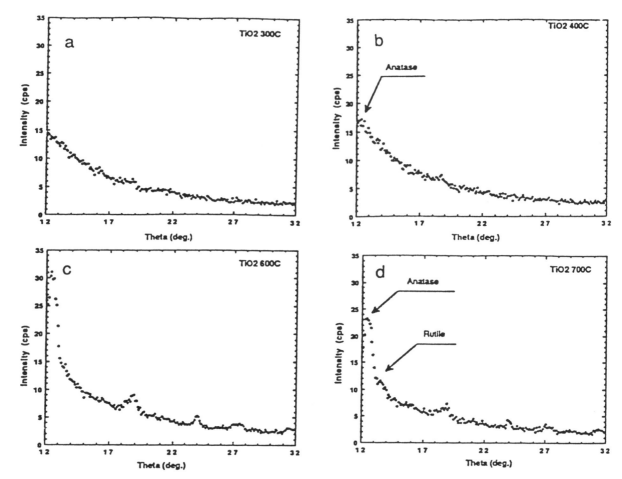

Fig. 3. XRD on 150 nm TiO$_2$ coatings dried at 60°C for 10 min and fired in air for 15 min at (a) 300°C, (b) 400°C, (c) 600°C, and (d) 700°C.

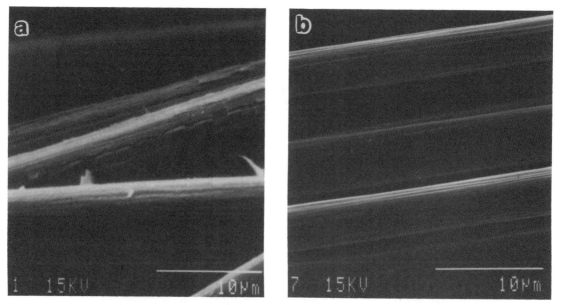

Fig. 4. PAN 650/42 fibers dipped twice in sol, dried at 60°C for 10 min after each dip, and fired at 700°C in (a) air, or (b) a CO atmosphere after final dip.

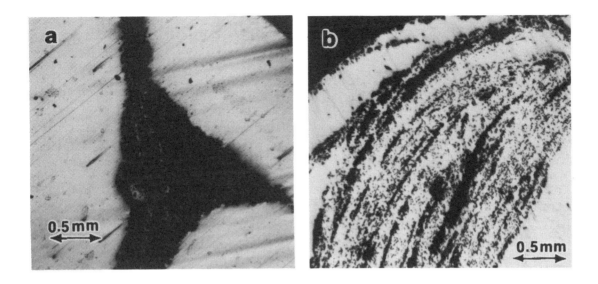

Fig. 5. Optical micrographs of pressure-infiltrated samples with (a) uncoated and (b) coated carbon fibers.

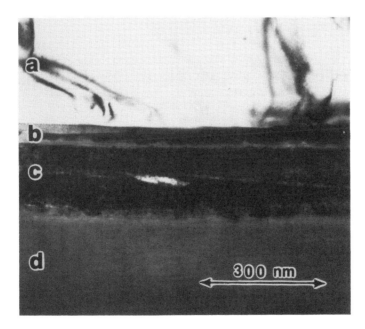

Fig. 6. TEM micrograph of the fiber-matrix interface; (a) aluminum, (b) reaction layer, (c) TiO_2 coating, and (d) carbon fiber.

REFERENCES

1.   Amateau, M. F., "Progress in the Development of Graphite Aluminum Composites Using Liquid Infiltration Technology," Jn. of Composite Materials, Vol. 10, Oct. 1976, pp. 279-296.

2.   Katzman, H. A., "Fibre Coatings for Fabrication of Graphite-Reinforced Magnesium Composites", Jn.Mat.Sci., Vol.22, 1987, pp. 144-148.

3.   Bennett, M. J., "Application and Evaluation of Ceramic Coatings Produced by Sol-Gel Technology and Vapor Deposition Procedures," Coating for High Temperature Application, E. Lang, ed., Applied Science Publishers, New York, 1983, pp. 169-192.

4.   Levitt, A., Di Cesare, E., and Wolf, S., Fabrication and Properties of Graphite Fiber Reinforced Magnesium, Army Materials and Mechanics Research Center; AMMRC TR 71-44; AD 735-313; Nov.1971.

5.   Cornie, J. A., Mortensen, A., and Fleming, M. C., "Wetting, Fluidity and Solidification in Metal Matrix Composite Castings: a Research Summary", Sixth International Conference on Composite Materials, F. L. Matthews, N. C. Buskell, J. M. Hodgkinson, and J. Morton, eds., Elsevier Applied Science, London, 1987, Vol.2, pp. 2.297-2.319.

6.   Cheng, J., and Wang, D., "Structural Transformation of the TiO_2-SiO_2 System Gel During Heat-Treatment", Jn. of Non-Crystalline Solid, Vol. 100, North-Holland, Amsterdam, 1988, pp. 288-291.

7.   Sherer, S., "Drying Gels II", Jn. of Non-Crystalline Solid, Vol. 89, North-Holland, Amsterdam, 1987, pp. 217-227.

8.   Mukherjee, S. P., "Inorganic Oxide Gels and Gel-Monoliths: their Crystallizatio Behavior", Emergent Process Methods for High-Technology Ceramics, R. F. Davis, H. Palmour, and R. L. Porter, eds., Plenum Press, NY, 1984, pp.95-109.

9.   Stull, D. R., and Prophet, H., JANAF Thermochemical Tables, second edition, National Standard Reference Data System, National Bureau of Standard publ., June 1972.

10.   Yoldas, B. E., "Deposition and Properties of Optical Oxide Coating from Polymerized Solutions", Applied Optic, Vol. 21, N°16, August 82, pp. 2960-2964.

11.   An, H. H., and Luhman, T. S., "Sol-Gel Coating Concepts for Very High Temperature Aluminium Matrix", Dispersion Strengthened Aluminium Alloys, Y. W. Kim, and W. M. Griffith, eds., The Mineral, Metals, and Materials Society, 1988, pp. 709-718.

12.   Chin, E. C., and Nunes, J., Alloying Effects in Graphite-Magnesium Composites, Presented at the TMS Annual Meeting in Phoenix, Arizona January 26, 1988, Unpublished.

Bimodal Viscoplasticity in Fibrous Metal-Matrix Composite Materials

R. HALL

ABSTRACT

A small strain theory of high—temperature, isothermal behavior for a fibrous metal—matrix composite lamina in plane stress is presented. The "matrix—dominated mode" (MDM) model proposed by Dvorak and Bahei—El—Din [1,2] is extended to rate—dependent conditions, using an adaptation of the viscoplasticity theory of Yen and Eisenberg [3]. The MDM formulation assumes that the plastic deformation in certain stress states is due only to matrix slip on planes containing the fiber axis, and is analogous to single—crystal plasticity. The growth law for the internal stress follows from consistency conditions for the resolved equilibrium stresses on the active planes, allowing the hardening direction to be specified. Anisotropies of the plastic flow parameters and yield function are included in the development.

MDM response is associated with the majority of the plane stress yield surface in boron/aluminum fibrous composites [1,2]; similar, MDM—dominant behavior is predicted for any metal—matrix system with the axial shear stiffness of its fibers much greater than that of its matrix [1].

INTRODUCTION

It was shown by Dvorak and Bahei—El—Din [1,2] that the yield surfaces of boron/aluminum unidirectional, thin tubes are accurately predicted by a "bimodal" plasticity theory. These yield surfaces are very different from the surfaces of elliptical cross—section which are predicted by volume—averaging approaches which assume a von Mises yield behavior in the matrix phase, e.g., [4]. The striking feature of the surfaces is that their periphery is dominated by segments where yield is dependent only on a critical value of axial shear stress, independent of axial and transverse normal stresses; that is, these regions appear flat (figures 1—2). It was shown in [1,2] that these flat "branches," and in fact, the entire crossection in the axial shear stress/transverse normal stress plane, are accurately predicted by assuming that yield is governed by a critical magnitude of resolved shear stress (effective) on planes containing the fiber axis. Deformation in this condition is thus assumed to occur by

Research Assistant, Dept. of Mechanical Engineering, Aeronautical Engineering and Mechanics, Rensselaer Polytechnic Institute, Troy, N.Y. 12180—3590 slip on the critically loaded planes, while slip on nonaxial planes is impeded by the fibers. A plasticity theory analogous to those employed for single crystals, e.g. [5–8], results; it is referred to as "matrix–dominated mode" (MDM), and is the subject of this paper. References listed in [1] indicate that similar deformation postulates have been investigated by previous authors.

The remaining portions of the yield surfaces can be predicted with a volume–averaging approach, with matrix yield governed by the same critical shear stress employed in the MDM criterion [1,2]. The associated deformation is referred to as "fiber–dominated mode" (FDM). The FDM model is currently under development and will not be discussed here, except to note that the FDM surfaces of figures 1–6 were calculated using the "self–consistent" method (SCM) [10]. It should be noted, additionally, that such a model is used, in lieu of measured elastic moduli, to calculate elastic strains in MDM.

Both MDM and FDM models are required for a "bimodal" characterization which is valid for all loading conditions; however, the MDM behavior asserts itself in such a dominant way in systems with the elastic, axial shear stiffness of the fibers much greater than that of the matrix (e.g., B/Al figures 1–2, from [2],and experimentally unverified SiC/Ti_3Al calculations, figures 3–6), that the MDM theory is potentially very useful in its own right. This can be particularly true in laminate systems; calculations analogous to those in ref. [9], but using the MDM yield criterion which follows, predict that a $(0/\pm45)_s$ SiC/Ti_3Al thin laminate is completely governed by MDM deformation in the stress planes of figures 5,6. (The internal envelope of the yield surfaces is entirely MDM.)

GENERAL CONSIDERATIONS, STRESS DEFINITIONS AND EQUILIBRIUM (YIELD) SURFACES

Following Yen and Eisenberg [3], it is assumed that there exists an equilibrium stress surface, or yield surface, centered at internal stress α, which is the boundary for inelastic behavior for all loading rates, including the equilibrium rate where the plastic strain rate $\dot{\epsilon}^p$ approaches zero. The equilibrium "loading rate" results in stress–strain behavior which can be modeled using standard rate–independent plasticity techniques, with the equilibrium stress point σ^* always lying on the equilibrium stress surface. The increments $d\epsilon^p$, $d\alpha$ and $d\sigma^*$ are thus taken to be related by rate–independent theory.

For nonequilibrium loading rate, the stress σ lies outside of the equilibrium stress surface, and the plastic strain rate is modeled as a function of the resolved shear overstress H on the actively slipping planes:

$$H \equiv \sqrt{(\tau_{ij} - \tau_{ij}^*)(\tau_{ij} - \tau_{ij}^*)} \equiv \sqrt{(\tau - \tau^*):(\tau - \tau^*)} \equiv \| \tau - \tau^* \| \tag{1}$$

where the components of τ are the resolved shear stress components on the active plane of interest, and their conjugates, and τ^* is the analogous resolved equilibrium shear stress corresponding to the same plastic strain history and equilibrium loading.

To find the stress components in (1), the coordinate systems of figures 7–8 are introduced. (Note that all vectors are identified with arrows, while bold–face identifies 2nd–rank tensors.) Define:

$$\vec{\tau} = \vec{n} \cdot \sigma - (\vec{n} \cdot \sigma \cdot \vec{n})\, \vec{n} \tag{2}$$

$$\tau_1 = \vec{\tau} \cdot \vec{e}_1 = \vec{n} \cdot \sigma \cdot \vec{e}_1 = \sigma_{n1} \tag{3}$$

$$\tau_\perp = \vec{\tau} \cdot \vec{e}_\perp = \vec{n} \cdot \sigma \cdot \vec{e}_\perp = \sigma_{n\perp} \tag{4}$$

$$\vec{n} = \cos \beta \, \vec{e}_2 - \sin \beta \, \vec{e}_3 \tag{5}$$

$$\vec{e}_\perp = \sin \beta \, \vec{e}_2 + \cos \beta \, \vec{e}_3 \tag{6}$$

$\vec{\tau}$ is the resolved shear stress vector on the axial plane with unit normal $\vec{n}$, and τ_1 and τ_2 are its components, as indicated in figure 8. $x_\perp$ is the coordinate direction, lying in the plane, which is orthogonal to the fiber direction x_1. $\vec{e}_\perp$ and $\vec{e}_1$ are unit vectors associated with these coordinates. The notation $\sigma_{n\perp}$, for example, indicates (eq. 4) the stress component, on the plane with normal direction $\vec{n}$, in the $\vec{e}_\perp$ direction.

In exactly the same manner, the internal stress α and equilibrium stress σ^* have corresponding resolved stress vectors $\vec{\alpha}_\tau$ and $\vec{\tau}^*$, with components $\alpha_1 = \alpha_{n1}$, $\alpha_\perp = \alpha_{n\perp}$ and $\tau_1^* = \sigma_{n1}^*$, $\tau_\perp^* = \sigma_{n\perp}^*$, respectively.

The equilibrium surface for a given plane can now be written:

$$\left(\tau_1^* - \alpha_1\right)^2 + \left(\tau_\perp^* - \alpha_\perp\right)^2 = \left(\tau_0 + R\right)^2 \tag{7}$$

where τ_0 is the initial critical shear stress, assumed here, as in [1,2], to be identical on every plane, and R is the plane–independent change in critical shear stress. (It is evident from figures 1–2 that the subsequent yield surfaces can be assumed to maintain the initial shape.) This surface is depicted in figure 9. In analogy to the definition of Yen and Eisenberg [3], the equilibrium stress $\vec{\tau}^*$ is defined as the intersection of the circle of radius $(\tau_0 + R)$ with the resolved effective stress $(\vec{\tau} - \vec{\alpha}_\tau)$, figure 9. Finally, the 2nd–rank tensors τ, τ^* and α_τ are introduced (to facilitate transformations and comparisons with constitutive laws for unreinforced materials), and the components of equation (1) are now identified:

$$\tau \equiv \tau_1 \left(\vec{n}\vec{e}_1 + \vec{e}_1\vec{n}\right) + \tau_\perp \left(\vec{n}\vec{e}_\perp + \vec{e}_\perp\vec{n}\right) \tag{8}$$

$$\tau^* \equiv \tau_1^* \left(\vec{n}\vec{e}_1 + \vec{e}_1\vec{n}\right) + \tau_\perp^* \left(\vec{n}\vec{e}_\perp + \vec{e}_\perp\vec{n}\right) \tag{9}$$

$$\alpha_\tau \equiv \alpha_1 \left(\vec{n}\vec{e}_1 + \vec{e}_1\vec{n}\right) + \alpha_\perp \left(\vec{n}\vec{e}_\perp + \vec{e}_\perp\vec{n}\right) \tag{10}$$

Equation (7) is now rewritten in terms of tensorial arguments:

$$f = \sqrt{(\tau^* - \alpha_\tau) : (\tau^* - \alpha_\tau)} - \sqrt{2} \left(\tau_0 + R\right) = 0 \tag{11}$$

According to the discussion following equation (7), the following can be written, defining the equilibrium shear stresses once σ, α and R are known:

$$\vec{\tau}^* - \vec{\alpha}_\tau = L \left(\vec{\tau} - \vec{\alpha}_\tau\right), \qquad\qquad \tau^* - \alpha_\tau = L \left(\tau - \alpha_\tau\right) \tag{12}$$

$$L = (\tau_0 + R) / \sqrt{(\tau_1 - \alpha_1)^2 + (\tau_\perp - \alpha_\perp)^2} \tag{13}$$

To obtain (12), the overall equilibrium stress satisfies (consider also eq. (19)):

$$\sigma^* - \boldsymbol{\alpha} = L\,(\sigma - \boldsymbol{\alpha}) \tag{14}$$

The overall equilibrium surface is pictured in figure 10. In general, the MDM deformation mode is active whenever σ^* lies on an MDM "branch"; otherwise, FDM mode is active. (Recall figures 1–6.)

To find the active slip planes, the (nonequilibrium) criterion of Dvorak and Bahei–El–Din [1,2] is, essentially, employed: slipping planes are those of maximum magnitude of resolved effective stress. From figure 9 and equations (12) and (1), this criterion is seen to produce planes of maximum overstress, H, as well (recall that R is plane–independent):

$$\frac{\partial}{\partial \beta}\,\|\tau - \boldsymbol{\alpha}_\tau\| = \frac{\partial}{\partial \beta}\,(H + \sqrt{2}(\tau_0 + R)) = 0 \tag{15}$$

The left–hand side of equation (15), together with (8), (10), (3)–(6) and the equations for α_1 and $\alpha_\perp$ analogous to (3)–(4), yields the following general (full stress space) relation for the active planes (this relation can also be obtained [11] using the approach in [1]):

$$\tfrac{1}{2}\,(\overline{\sigma}_{21}^{2} - \overline{\sigma}_{31}^{2})\sin 2\beta + \overline{\sigma}_{21}\overline{\sigma}_{31}\cos 2\beta$$

$$+ \,[\overline{\sigma}_{32}^{2} - ((\overline{\sigma}_{33} - \overline{\sigma}_{22})/2)^2]\sin 4\beta + \overline{\sigma}_{32}(\overline{\sigma}_{33} - \overline{\sigma}_{22})\cos 4\beta = 0 \tag{16a}$$

$$\text{where } \overline{\sigma}_{ij} \equiv \sigma_{ij} - \alpha_{ij}. \tag{16b}$$

Note that $\overline{\sigma}$ can be replaced in (16) by any scalar multiple of $\overline{\sigma}$, including $(\sigma^* - \boldsymbol{\alpha})$ (see eq. (14)), and the same equation for the slip planes results.

It has been shown by the author [12] that, for general loading, equation (16) possesses at most two solutions β_i (i=1,2) which define unique slip planes. For general loading, these solutions are found either by iteration or by squaring (16) and finding the appropriate roots of the resulting quartic polynomial in $\cos 2\beta$. Henceforth, a single index "i" will be used to identify quantities associated with the active plane "i" (i=1,2). The usual summing conventions will not apply to this index, and no summing conventions will be employed in the sequel.

PLANE STRESS EQUILIBRIUM (YIELD) SURFACE

With σ_{11}, σ_{22}, and σ_{21} the only nonzero applied stresses, the two unique solutions of (16) can be taken as:

$$\beta_{1,2} = \pm\,\cos^{-1}\sqrt{(1 + q^2)/2} \qquad\qquad |q| \le 1 \tag{17a}$$

$$\beta_{1,2} = 0 \qquad\qquad\qquad\qquad\qquad\qquad |q| \ge 1 \tag{17b}$$

$$q \equiv (\sigma_{21} - \alpha_{21})/(\sigma_{22} - \alpha_{22}) = (\sigma_{21}^* - \alpha_{21})/(\sigma_{22}^* - \alpha_{22}) \tag{18}$$

Equations (17–18) can be used with equations (3)–(7) and the corresponding α_1, $\alpha_\perp$ relations to produce the relations of reference [2] for the plane stress yield surface under equilibrium loading:

$$(\overline{\sigma}^*_{21}/\overline{\tau})^2 + ((\overline{\sigma}^*_{22}/\overline{\tau})\pm1)^2 = 1 \qquad\qquad |q|\leq1 \qquad\qquad (19a)$$

$$(\overline{\sigma}^*_{21}/\overline{\tau})^2 = 1 \qquad\qquad |q|\geq1 \qquad\qquad (19b)$$

$$\text{where } \overline{\sigma}^*_{ij} \equiv \sigma^*_{ij}-\alpha^*_{ij}, \ \overline{\tau} \equiv \tau_0 + R \qquad\qquad (19c)$$

Equations (19) are used to construct the MDM branches of the single lamina yield surfaces of figures 1–6. The complete surface is the internal envelope of the MDM and FDM surfaces.

FLOW

The standard rate–independent plasticity equations give the equilibrium flow rules on the two active planes, i = 1,2:

$$d\epsilon^p_i = \mathbf{N}_i \, (\mathbf{N}_i{:}d\boldsymbol{\sigma}^*)/h_i \qquad\qquad (20a)$$

$$\mathbf{N}_i = \frac{\partial f_i}{\partial \sigma^*} = \frac{\tau^* - \boldsymbol{\alpha}_\tau}{\|\tau^* - \boldsymbol{\alpha}_\tau\|} = \frac{\tau - \boldsymbol{\alpha}_\tau}{\|\tau - \boldsymbol{\alpha}_\tau\|} \qquad\qquad (20b)$$

where (20b) follows from differentiating (11).

Define the slip vector $\vec{s}_i$, oriented at angle θ_i from the $\vec{e}_1$–direction, as indicated in figure 9 (and in figure 8 for initial yield):

$$\vec{s}_i(\theta_i) \equiv \frac{(\tau_1 - \alpha_1)\vec{e}_1 + (\tau_\perp - \alpha_\perp)\vec{e}_\perp}{\sqrt{(\tau_1 - \alpha_1)^2 + (\tau_\perp - \alpha_\perp)^2}}\bigg|_i \qquad\qquad (21)$$

Equation (20b) can now be written, using (8), (10) and (21):

$$\mathbf{N}_i = (\vec{n}_i\vec{s}_i + \vec{s}_i\vec{n}_i)/\sqrt{2} \qquad\qquad (22)$$

The rate–dependent flow rules for the active planes are taken here, for simplicity, in the power–law form:

$$d\epsilon^p_i = \mathbf{N}_i \, (H/K_i)^{n_i} \, dt \qquad\qquad (23)$$

where H is identical on the two active planes.

As in [3], it is noted that the stresses in equations (20) and (23) correspond to the same plastic strain history, so (20a) and (23) can be equated, producing the relation:

$$\mathbf{N}_i{:}\dot{\boldsymbol{\sigma}}^* = h_i(H/K_i)^{n_i} \qquad\qquad (24)$$

The total plastic strain rate is given by the sum of the contributions from the two active planes, and is calculated from (23):

$$\dot{\epsilon}^p = \dot{\epsilon}^p_1 + \dot{\epsilon}^p_2 \qquad\qquad (25)$$

It is shown in [12] that the rate–independent results presented here can be used to recover all of the relations of Dvorak and Bahei–El–Din [1].

SYMMETRY CONDITIONS

From (22), with (21), (16),(3)–(6) and the discussions following (16) and (6), it follows that:

$$\mathbf{N}_i = \mathbf{N}_i(\beta_i(\mathbf{y}), \theta_i(\mathbf{y})) \quad , \quad \mathbf{y} \equiv (\sigma-\alpha)/\|\sigma-\alpha\| \tag{26}$$

Since the plastic strain rate ratios embodied in $\mathbf{N}_i(\mathbf{y})$ are the same, for a given value of $\mathbf{y}$, as they were in the initial state, the normalized effective stress $\mathbf{y}$ always acts on a material possessing the initial symmetry.

The plastic flow parameters h_i, K_i, n_i must depend on the flow direction $\mathbf{N}_i$ and a set of scalars. The dependence on $\mathbf{N}_i$ can be expressed, from (26), as dependence on the initial material angles β_i and θ_i. The directional dependence of the flow parameters is thus reduced, through rotational symmetry about, and reflective symmetry through, the fiber direction $\vec{e}_1$, to dependence on the acute angle $\hat{\theta}_i$ which $\vec{s}_i$ makes with $\vec{e}_1$. The flow parameter dependence can now be written:

$$h_i = h(\hat{\theta}_i(\mathbf{y}), \phi_h) \tag{27a}$$

$$K_i = K(\hat{\theta}_i(\mathbf{y}), \phi_K) \tag{27b}$$

$$n_i = n(\hat{\theta}_i(\mathbf{y}), \phi_n) \tag{27c}$$

where ϕ_h, ϕ_K, ϕ_n represent three sets of scalars.

FLOW AND KINEMATIC HARDENING IN PLANE STRESS

From the definition of θ_i previous to (21), $\tan \theta_i = [(\tau_2-\alpha_2)/(\tau_1-\alpha_1)]_i$; using (3)–(6) and the associated equations for α_1 and $\alpha_\perp$ it is found, for σ_{11}, σ_{22} and σ_{21} the only nonzero applied stresses, that:

$$|\tan \theta_i| = |\tan \hat{\theta}_i| = |(\sin \beta_i)/q| \tag{28}$$

From (28) and (17) it follows that $\hat{\theta}_1 = \hat{\theta}_2$. Equations (27) then demand that the flow parameters h, K and n are identical, in plane stress, on the two active planes.(The subscripts associated with these parameters are, henceforth, omitted.)

The kinematic hardening rule is determined, as in [3], from the consistency conditions for the equilibrium stress. Considering either active plane:

$$\dot{f}_i = 0 = \frac{\partial f_i}{\partial \sigma} : \dot{\sigma}^* + \frac{\partial f_i}{\partial \alpha} : \dot{\alpha} + \frac{\partial f_i}{\partial R} \dot{R} \tag{29}$$

$$\dot{\alpha} \equiv \dot{\mu}\nu \tag{30}$$

Equations (29), (30), (11) and (20b) yield:

$$\dot{\mu} = (\mathbf{N}_i:\dot{\sigma}^* - \sqrt{2}\dot{R})/\mathbf{N}_i:\nu \tag{31}$$

For plane stress, it can be verified that the in–plane (i.e., 11, 22, 21) components of both N_i are identical, and the out–of–plane components are of equal magnitude and opposite sign. The terms in (31) are, therefore, identical for both planes, and $\dot{\epsilon}_p$ has no out–of–plane shearing components from (20a), (23), (25) and the discussion following (28). Equation (24) completes the rule for $\dot{\mu}$:

$$\dot{\mu} = (h(H/K)^n - \sqrt{2}\dot{R})/N_i{:}\nu \tag{32}$$

The hardening direction ν is left arbitrary, in order to allow matching with experiment. For loading on the curved branches of the yield surface, the boron/aluminum experiments of [2] showed very accurate agreement with the stress–rate, or Phillips [13], direction. (For such materials, the equilibrium, rate–independent character of eq. (30) suggests that the direction of ν in rate–dependent loading is the equilibrium stress–rate, not the stress–rate, direction; measured, near–equilibrium yield surface motions are still recovered. Experiments of Phillips [14] and others, e.g., [15], on unreinforced materials, indicate that the yield surface motion does not match (lags behind) the loading point motion for nonequilibrium rates, in accordance with the model employed here and in [3] and [16], for example. If the stress–rate is held at zero following nonequilibrium loading, the yield surface moves, in creep, toward the loading point [14].)

For loading on the flat branch of the yield surface, neutral loading–type arguments [12] predict that the only possible yield surface motion is in the direction normal to the surface. This behavior was observed in [2].

At high homologous temperature, the internal stress is allowed to recover according to an established format [16,17]:

$$\dot{\alpha} = \dot{\mu}\nu - a\,\|\alpha\|^u\,\alpha \tag{33}$$

The first term corresponds to the hardening rate in the absence of thermal recovery.

SCALAR HARDENING VARIABLES

The equilibrium plastic modulus h is characterized with a bounding surface formulation in the $\tau_1^*/\tau_\perp^*$ plane, in a way analogous to , e.g., [18,19], but with $\hat{\theta}$–dependent parameters. In the plane stress state previously discussed, all stresses in either of the active planes are mirrored in the other [12], so either plane can be used to determine h. Space limitations preclude further details, but it is noted that an inital value for h which is essentially infinite will lead to numerical problems from equation (32). Treatments such as [18] must be modified to incorporate a suitable value for the initial modulus.

The change in critical stress, R, is currently allowed to evolve according to standard [16,17] formats, and includes thermal recovery:

$$\dot{R} = b(R_1 - R)\dot{M} - c(R - R_2)^m \tag{34}$$

$$\dot{M} = \|\dot{\epsilon}^p\| \text{ or } \sigma{:}\dot{\epsilon}^p \tag{35}$$

The proposed evolution of drag stress, K, is similar, but is currently allowed to depend on $\hat{\theta}$ in the following way:

$$K(\hat{\theta}) = r(\hat{\theta})K_M + K_T \tag{36a}$$

$$\dot{K}_M = r_1(\hat{\theta})(K_1(\hat{\theta}) - K_M)\dot{M} \tag{36b}$$

$$\dot{K}_T = -A(K_M - \overline{K}_0)^S \tag{36c}$$

Experience will dictate whether or not expressions (34)–(36) need to be modified, or are even, perhaps, unnecessarily general.

ACKNOWLEDGEMENTS

The author is grateful to Drs. G.J. Dvorak, Y.A. Bahei–El–Din, E. Krempl and K.P. Walker for useful discussions, encouragement and help during the literature search. Thanks also to R. Shah for assisting with figures 5–6. The financial support for this work was provided by ONR Defense Advanced Research Projects Agency (DARPA) Contract no. N00014–86–K–0770 and is gratefully acknowledged.

REFERENCES

1 Dvorak, G.J., and Bahei-El-Din, Y.A., "A Bimodal Plasticity Theory of Fibrous Composite Materials," Acta Mech., Vol. 69, 1987, pp. 219-241.

2 Dvorak, G.J., Bahei-El-Din, Y.A., Macheret, Y., and Liu, C.H., "An Experimental Study of Elastic-Plastic Behavior of a Fibrous Boron-Aluminum Composite," J. Mech. Phys. Solids, Vol 36, 1988, pp. 655-687.

3 Yen, C.F., and Eisenberg, M.A., "The Role of a Loading Surface in Viscoplasticity Theory," Acta Mech., Vol. 69, 1987, pp. 77-96.

4 Dvorak, G.J., and Bahei-El-Din, Y.A., "Plasticity Analysis of Fibrous Composites," J. Appl. Mech., Vol. 49, 1982, pp. 327-335.

5 Asaro, R.J., Micromechanics of Crystals and Polycrystals, Adv. in Applied Mechanics, Vol. 23, 1983, pp. 1-115.

6 Lin, T.H., Physical Theory of Plasticity, Adv. in Applied Mechanics, Vol. 11, 1971, pp. 255-311.

7 Dame, L.T., Anisotropic Constitutive Model for Nickel Base Single Crystal Alloys: Development and Finite Element Implementation, Dissertation, University of Cincinnati, 1985.

8 Walker, K.P., and Jordan, E.H., "Biaxial Constitutive Modelling and Testing of a Single Crystal Superalloy at Elevated Temperatures," Biaxial and Multiaxial Fatigue, EGF3, Mechanical Engineering Publications, London, 1988, pp. 145-170.

9 Bahei-El-Din, Y.A., and Dvorak, G.J. "Plasticity Analysis of Laminated Composite Plates," J. Appl. Mech., Vol. 49, 1982, pp. 740-746.

10 Walpole, L.J., "On the Overall Elastic Moduli of Composite Materials," J. Mech. Phys. Solids, Vol. 17, 1969, pp. 235-251.

11 Dvorak, G.J., and Bahei-El-Din, Y.A., "New Results in Bimodal Plasticity of Fibrous Composite Materials," to be published.

12 Hall, R., Dissertation(in progress), Dept. of Mechanical Engineering, Aeronautical Engineering and Mechanics, Rensselaer Polytechnic Institute, 1989.

13 Phillips, A., and Weng, G.J., "An Analytical Study of an Experimentally Verified Hardening Law," J. Appl. Mech., Vol. 42, 1975, pp. 375-378.

14 Phillips, A., "A Review of Quasistatic Experimental Plasticity and Viscoplasticity," Int. J. Plasticity, Vol. 2, 1986, pp. 315-328.

15 Moreton, D.N., Moffat, D.G., and Parkinson, D.B. "The Yield Surface Behavior of Pressure Vessel Steels," J. Strain Analysis, Vol. 16, 1981, pp. 127-136.

16 Chaboche, J.L., "Cyclic Plasticity Modeling and Ratchetting Effects," Constitutive Laws for Engineering Materials: Theory and Applications, Elsevier, New York, 1987, pp. 1165-1172.

17 Lindholm, U.S., Chan, K.S., Bodner, S.R., Weber, R.M., Walker, K.P., and Cassenti, B.N., "Constitutive Modeling for Isotropic Materials (HOST)," NASA CR-174718, 1984.

18 Dafalias, Y.F., and Popov, E.P., "Plastic Internal Variables Formalism of Cyclic Plasticity," J. Appl. Mech., Vol. 43, 1976, pp. 645-651.

19 McDowell, D.L., "Simple Experimentally Motivated Cyclic Plasticity Model," J. Eng. Mech., Vol. 113, 1987, pp. 378-397.

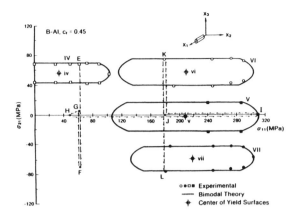

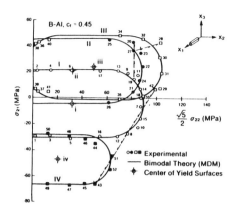

Figure 1. Axial tension/axial shear yield surfaces for a uni-directional B/Al tube. From [2].

Figure 2. Transverse tension/axial shear yield surfaces for a uni-directional B/Al tube. From [2].

SiC/Ti₃Al with SCM FDM

SiC/Ti₃Al with SCM FDM

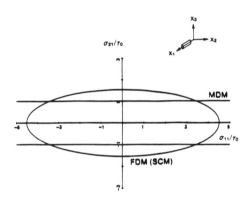

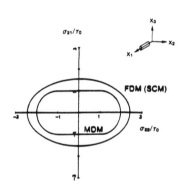

Figure 3. Calculated axial tension/axial shear yield surface for a SiC/Ti₃Al lamina. Fiber volume fraction 0.5.

Figure 4. Same as fig. 3, but transverse tension/axial shear loading.

SiC/Ti₃Al (0/±45)ₛ LAMINATE, cᵣ=0.5

SiC/Ti₃Al (0/±45)ₛ LAMINATE, cᵣ=0.5

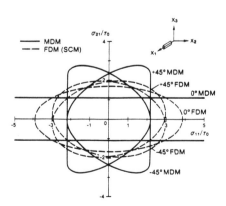

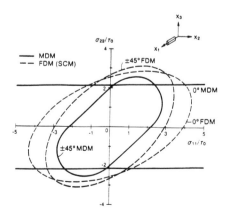

Figure 5. Same as fig. 3, but for a (0/±45)ₛ laminate.

Figure 6. Same as fig. 4, but for a (0/±45)ₛ laminate.

29

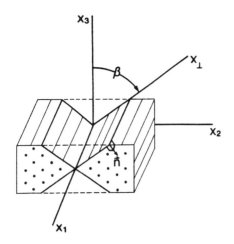

Figure 7. Lamina coordinate system.

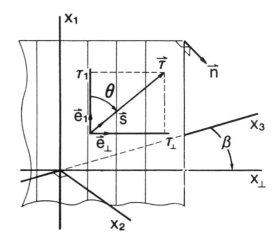

Figure 8. Slip plane coordinate system.

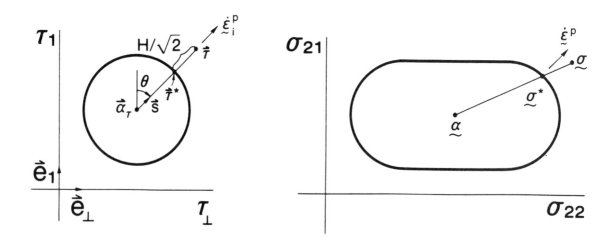

Figure 9. Slip plane equilibrium surface.

Figure 10. Lamina equilibrium surface.

Guided Waves and Defect Scattering in Metal Matrix Composite Plates

SUBHENDU K. DATTA AND
ROBERT L. BRATTON

ARVIND H. SHAH

ABSTRACT

Guided Rayleigh-Lamb waves in a continuous graphite fiber reinforced magnesium plate has been studied in this paper. The interest in this material arises from its high thermal stability and because it provides high strength to weight ratio. Our previous studies have shown that for wave lengths much larger than the fiber diameters and spacing the material can be characterized as transversely isotropic with the symmetry axis aligned with the fiber direction. It is also found that because of the high longitudinal stiffness of the graphite fibers the material shows strong anisotropy with very high modulus in the fiber direction. For this reason it is found that dispersion of guided waves is strongly influenced by the deviation of the direction of propagation from the symmetry axis. In this paper we present results for propagation in different directions. In addition, we present results for scattering of anti-plane shear waves by surface breaking cracks and delaminations when the propagation is in the fiber direction.

INTRODUCTION

There is currently considerable interest in metal matrix composites for applications in space structures. Both particle and fiber reinforced materials are under investigation. Our recent studies [1,2] have shown that these materials can usually be characterized as transversely isotropic having five distinct elastic stiffnesses. Using a wave scattering formalism, models of their rheology were derived for predicting these five elastic stiffnesses. Manufactured parts (plates, tubes, etc.) containing these materials have unique properties, which are subjects of considerable interest for ultrasonic nondestructive evaluations, impact response, and vibrations. In this paper we have studied guided wave propagation in plates made of graphite fiber-reinforced magnesium. As was shown in previous investigations [2], this material shows transverse isotropic symmetry. Here it has been assumed that the axis of symmetry lies in the plane of the plate. Thus for propagation in an arbitrary direction parallel to the plate, the motion is three dimensional, i.e., the equations governing the three components of displacement are coupled. This causes considerable complexity in the dispersion equation. Here we have presented solutions to this equation showing different behaviors for propagation in different directions.

Subhendu K. Datta, Professor, Robert L. Bratton, Research Assistant, Department of Mechanical Engineering and Center for Space Construction, University of Colorado, Boulder, CO 80309-0427; Arvind H. Shah, Professor, Department of Civil Engineering, University of Manitoba, Winnipeg, Canada R3T 2N2

Wave propagation in homogeneous isotropic plates and monoclinic plates have been studied extensively using exact and approximate methods by Mindlin and coworkers [3-6]. Recently, harmonic wave propagation in fiber reinforced composites has been the focus of investigations using both numerical and analytical techniques. Representative of the numerical methods is the stiffness method discussed in [7-10]. Analytical models for leaky Lamb waves in a fluid coupled composite plate have been presented in [11-12]. In the following section we derive the dispersion equation by solving analytically the differential equations governing the displacement components and satisfying the boundary conditions for a free-free plate.

GOVERNING EQUATIONS FOR RAYLEIGH-LAMB WAVES

Time harmonic waves of the form shown in Eq. (1) below are considered to represent waves guided by the plate, where ω represents the circular frequency (rad/sec), $\underset{\sim}{K}$ is the wave vector, $\underset{\sim}{u}$ (z) is the displacement, and $\underset{\sim}{f}$ (z) is an undertermined function of z.

$$\underset{\sim}{u} = \underset{\sim}{f}(z)e^{i\left[\underset{\sim}{K}\cdot\underset{\sim}{x} - \omega t\right]} \tag{1}$$

As shown in Figure 1, the wave propagation is at an angle ϕ with the x-axis. Hence $\underset{\sim}{K}$ is composed of x and y directional components which will be designated as K and ℓ, respectively. Thus $k = |K|\cos\phi$ and $\ell = |K|\sin\phi$, and

$$\underset{\sim}{u} = \underset{\sim}{f}(z)e^{i(kx+\ell y-\omega t)} \tag{2}$$

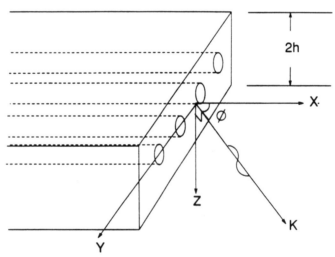

Fig. 1. Uniaxial fiber reinforced plate. $\underset{\sim}{K}$ indicates direction of wave propagation.

Using the method discussed in [13], $\underset{\sim}{u}$ is represented in terms of the pseudo-potentials, θ, Φ, ψ in the following form.

$$u_x = \frac{\partial\theta}{\partial x}, \ u_y = \frac{\partial\Phi}{\partial y} + \frac{\partial\psi}{\partial z}, \ u_z = \frac{\partial\Phi}{\partial z} - \frac{\partial\psi}{\partial y} \tag{3}$$

where,

$$\theta = f_1(z)e^{i(kx+\ell y-\omega t)} \qquad (4a)$$

$$\Phi = f_2(z)e^{i(kx+\ell y-\omega t)} \qquad (4b)$$

$$\psi = f_3(z)e^{i(kx+\ell y-\omega t)} \qquad (4c)$$

As seen in Fig. 1, the x-axis conincides with the axis of symmetry. The z-axis is in the direction of the thickness of the plate. The origin of the co-ordinates is taken in the mid-plane of the plate, which is assumed to be of thickness, 2h. The y-axis is out of the plane of the diagram. Using the constitutive relation for the transversely isotropic materials considered in this paper, the equations of motion, and Eqs. (3-4), it can be shown that the functions f_1, f_2 anf f_3 are given by

$$f_1(z) = F_1\Omega_1^+ + G_1\Omega_2^+$$

$$f_2(z) = F_2\Omega_1^+ + G_2\Omega_2^+ \qquad (5)$$

$$f_3(z) = F_3\Omega_3^-$$

where

$$\Omega_1^+ = A_{11} \cos s_1 z + A_{12} \sin s_1 z$$

$$\Omega_2^+ = A_{21} \cos s_2 z + A_{22} \sin s_2 z \qquad (6)$$

$$\Omega_3^- = A_{32} \cos rz - A_{31} \sin rz$$

The constants F_1, F_2 and G_1, G_2 can be chosen such that

$$\frac{F_1}{F_2} = -\frac{\delta(s^2_1+\ell^2)}{s^2_1+\ell^2+\alpha k^2-K^2_2} \qquad (7a)$$

$$\frac{G_1}{G_2} = -\frac{\beta(s^2_2+\ell^2)+k^2-K^2_2}{\delta k^2} \qquad (7b)$$

Note that s^2_1 and s^2_2 correspond to the + and − signs, respectively, in the expression for s^2 given below.

$$s^2 = \frac{\left\{[K_2^2(1+\beta)-k^2\gamma] \pm \sqrt{[k^2\gamma-K^2_2(1+\beta)]^2-4\beta(\alpha k^2-K^2_2)(k^2-K^2_2)}\right\}}{2\beta} - \ell^2 \qquad (8)$$

Similarly, using Eq. (4c) and the expressions for $f_3(z)$ given above we get r to be

$$r = \left[[K^2_2-k^2-\ell^2\epsilon]/\epsilon\right]^{1/2} \qquad (9)$$

We have defined $\alpha = C_{11}/C_{55}$, $\beta = C_{33}/C_{55}$, $\delta = C_{13}/C_{55} + 1$, $\gamma = 1+\alpha\beta-\delta^2$, $\epsilon = C_{44}/C_{55}$ and $K^2_2 = \rho\omega^2/C_{55}$.

With the expressions for θ, Φ, and ψ now completely determined, the traction components σ_{xz}, σ_{yz}, and σ_{zz} at the surfaces of the plate are set to zero. This leads to six linear homogeneous equations in the constants A_{11}, A_{21}, A_{31}, A_{12}, A_{22}, A_{32}. these equaitons can be grouped into a set of three equations with the first three constants corresponding to the symmetric motion and another set containing the second three corresponding to the anti-symmetric motion. Note that the constants F_1, G_1. amd F_3 may be taken as 1, and the other two constants F_2 and G_2 are them found from Eqs. (7a,b).

Thus, we obtain the dispersions equations for the symmetric and anti-symmetric modes of the plate as

$$\text{SYM:} \quad \left[\frac{E_{31}\Lambda_1}{E_{33}\Lambda_2} + \frac{\tan s_1 h}{\tan s_2 h} \right] \sin rh + \frac{E_{35}}{E_{33}\Lambda_2} \cos rh \tan s_1 h = 0 \qquad (10a)$$

$$\text{ANTI-SYM:} \quad \left[\frac{E_{33}\Lambda_2}{E_{31}\Lambda_1} + \frac{\tan s_2 h}{\tan s_1 h} \right] \cos rh + \frac{E_{35}}{E_{31}\Lambda_1} \frac{\sin rh}{\tan s_2 h} = 0 \qquad (10b)$$

$$\text{with} \quad \Lambda_1 = \frac{E_{44}E_{66} - E_{64}E_{46}}{E_{42}E_{64} - E_{62}E_{44}} \quad \text{and} \quad \Lambda_2 = \frac{E_{46}E_{62} - E_{42}E_{66}}{E_{42}E_{64} - E_{62}E_{44}}.$$

where the exprerssion for E's are omitted here. These two equations are solved for the non-dimensional frequency, Ω , as a function of the wavenumber ξ, which are define as

$$\Omega = \frac{\omega h}{2\pi \sqrt{C_{55}/\rho}} \;, \quad \xi = \frac{Kh}{2\pi} \;.$$

Once these roots are known the phase velocities are then calculated from the relation

$$C = \left[\frac{\Omega}{\xi} \right] \sqrt{C_{55}/\rho}.$$

Now for propagation along the fibers it can be shown that the equation governing u_y uncouples from those governing u_x and u_z. Thus in this case it is possible to study the anti-plane sheer ($u_y \neq 0$, $u_x = u_z = 0$) wave motion independently of the in-plane (plane strain) motion ($u_y = 0$, $u_x \neq 0$, $u_z \neq 0$). Furthermore, if the anti-plane sheer wave interacts with either a long surface breaking crack or a delmaination with the long dimension parallel to the y-axis then the scattered field is also anti-plane shear. This is particularly attractive for ultrasonic characterization of such defects. In the following we briefly outline the solution technqiue for scattering of anti-plane shear (SH) waves by such defects.

SCATTERING OF SH WAVES

The plate with either a normal surface breaking crack or a delamination defect (horizontal buried crack) is shown in Fig. 2.

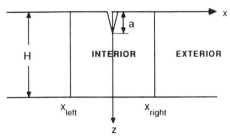

Fig. 2a. A normal surface breaking crack.

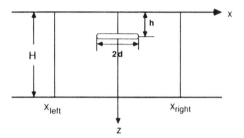

Fig. 2b. A buried horizontal crack (delamination).

Because SH waves are simpler to detect and analyze, here we will consider scattering of such waves by defects as shown in Fig. 2. The displacement u_y satisfying the equation of motion and boundary conditions at the free surfaces of the plate can be written as

$$u_y = A_m V_m^{(z)} e^{ik_m x} \tag{11}$$

for propagation along the positive x-axis. Here

$$V_m^{(z)} = \cos \frac{m\pi z}{H}$$

$$\text{and} \quad k_m = \sqrt{K_2^2 - \epsilon^2 \frac{m^2\pi^2}{H^2}}$$

with $K_2^2 = \dfrac{\rho\omega}{C_{44}}$, $\epsilon^2 = \dfrac{C_{66}}{C_{44}}$, and $m = 0,1,2,3,\ldots$.

We will examine the scattering of an incident wave in the mode m. The scattered field will be composed of all the modes and can be written as

$$u_y^{s^+} = \sum_{n=0}^{\infty} B_n^+ V_n^{(z)} e^{ik_n x} \ , \ x \geq x_{right} \tag{12a}$$

$$u_y^{s^-} = \sum_{n=0}^{\infty} B_n^- V_n^{(z)} e^{-ik_n x} \ , \ x \leq x_{left} \tag{12b}$$

The field in the interior region (see Fig. 2) is represented by finite elements and matching the displacements and tractions at the right and left vertical boundaries, x_{left} and x_{right}, we obtain equations for the determination of B_n^+ and B_n^-. We now define the reflection coefficient of the nth mode when the incident mode is m as

$$R_{mn} = \frac{B_n}{A_m} \tag{13a}$$

and the transmission coefficent as

$$T_{mm} = 1 + \frac{B_m^+}{A_m} \, , \quad T_{mn} = \frac{B_n^+}{A_m} \, , \quad n \neq m \tag{13b}$$

In the following section we present numerical results showing the dispersion of guided waves and the variation of the reflection and transmission coefficients with frequency for the two types of defects shown in Fig. 2.

DISCUSSION OF RESULTS

The constants for the Gr/Mg material are shown in Table 1. Gr/Mg material is a continuous fiber-reinforced composite, 60% by vol of graphite. To show the degree of anisotropy the bulk material velocity diagram is shown in Fig. 3. In this figure the dash-dot lines represent the quasi-longitudinal wave, the dashed lines represent the quasi-shear wave, the solid lines represent the anti-plane shear waves and the three dot-dashed lines represent the Rayleigh wave velocity. It is seen from Fig. 3, Gr/Mg material's large longitudinal wave velocity in the fiber direction is characteristic of substantial increase in the elastic stiffness in this direction over the other two directions. In the velocity diagrams the ordinate represents the velocity at 90^0 and the abscissa represents the velocity at 0^0 to the fibers.

TABLE 1

All stiffnesses are in units of 10^{11} N/m^2

Materials	C_{11}	C_{33}	C_{13}	C_{44}	C_{55}
Gr/Mg	1.806	0.273	0.0956	0.0738	0.219

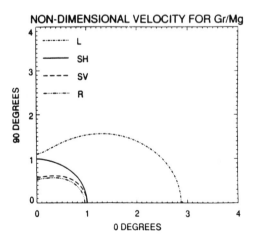

Fig. 3. Velocity Diagrams for Gr/Mg.

Consider now the phase velocity-frequency diagrams shown in Figs. 4 and 5. In Fig. 4 the normalized phase velocities for waves propagating parallel to the axis in the Gr/Mg composite plate are shown for the symmetric branches. These graphs depict the dispersion behavior of guided waves in the plate. The dispersion Eqs. (10a) and (10b) do not give solutions for phase velocities independent of frequencies, except for the first anti-plane shear wave propagating along the fiber and perpendicular to the fiber. The 0th mode anti-plane shear velocity is unity in both Figs. 4 and 5 since it is normalized with respect to $\sqrt{C_{55}/\rho}$. Figure 5 shows the dispersion behavior in the Gr/Mg plate for propagation at 45^0 to the fibers. The strong anisotropy of this material is seen to alter the dispersion characteristics drastically. It is found from Figs. 4 and 5 that the velocity curves for the symmetric modes show two distinct plateaus: one at the longitudinal velocity

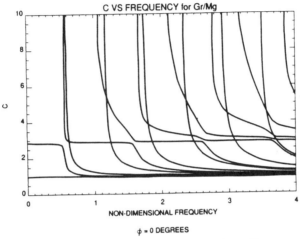

Fig. 4. Phase velocity for symmetric modes of propagation along the fiber direction in the Gr/Mg plate.

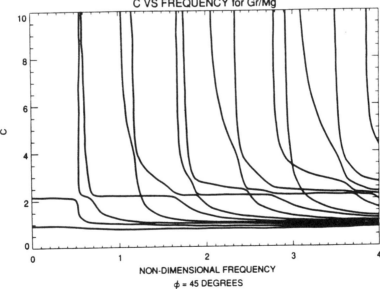

Fig. 5. Phase velocity for symmetric modes propagating at 45^0 to the fibers in the Gr/Mg plate.

in the plate and the other at the Rayleigh velocity (reached by the lowest modes) or the shear velocity (reached by the higher modes). The feature of (quasi) in-plane and (quasi) out-of-plane modes for off-axis propagation coming close to one another is strongly pronounced here also.

Figures 6 and 7 show the disparison curves for the antisymmetric modes. These figures also show the strong anisotropy effects.

In the following we present results of scattering of a particular propagating mode. Figure 8 shows the variation of reflection coefficients with frequency when the 0th mode wave is incident on a normal surface breaking crack. Here R[0,n] represents the reflection coefficient of the nth scattered mode when the incident mode is zero. It is seen from this figure that the reflection coefficients have slope discontinuities

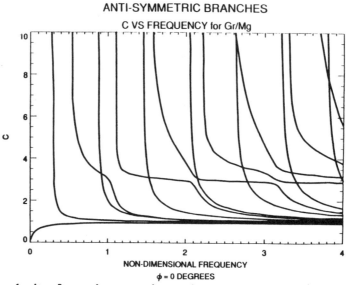

Fig. 6. Phase velocity for anisymmetric modes propagating along the fiber direction.

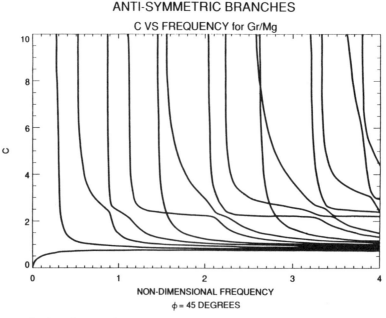

Fig. 7. Phase velocity for antisymmetric modes propagating 45° to the fiber direction.

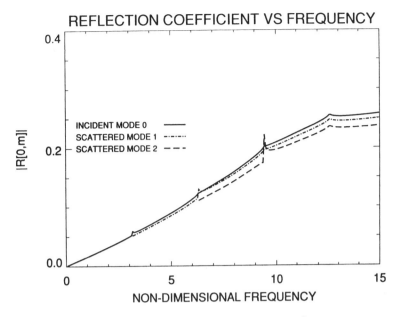

Fig. 8. Reflection coefficient of different modes when the 0th mode is incident.

at the cut-off frequencies of the successive modes. The figure shows that the reflection coefficients increase with frequency. The length of the crack is taken to be a = 0.15H. Figures 9 and 10 show the reflection coefficients for a buried horizontal crack (delamination). The crack is buried at a depth 0.15H and its length 0.2H. These figures show rather interesting resonance peaks at the cut-off frequencies. These peaks are particularly pronounced in the reflection coefficients of the second (n=1) and higher modes. It is seen from these figures that the detectability of cracks is enhanced only for a small band of frequencies around the cut-off values. Beyond these the reflection coefficients are very low and thus would not be useful for crack characterization.

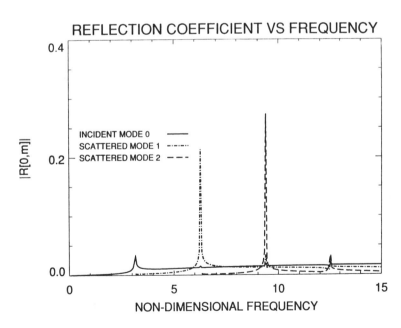

Fig. 9. Reflection coefficient of various modes for the first mode (n=0) incidence.

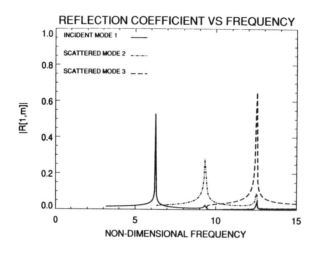

Fig. 10. Reflection coefficient of various modes for the 2nd mode (n=1) incidence.

ACKNOWLEDGEMENT

The work reported here was performed under the auspicies of the Center for Space Construction and has been supported in part by a grant from the Office of Naval Research (00014-86-K-0280) and by a grant from NASA (NAGW-1388). Partial support was also received from the National Science Foundation (MSM-8609813, INT-8521422, INT-8610487). The work of AHS was supported by a grant from the Natural Science and Engineering Research Council of Canada (A-7988).

REFERENCES

1.   Ledbetter, H. M. and Datta, S. K. "Effective Wave Speeds in an SiC-particle-reinforced Al Composite," Journal of the Acoustical Society of America, Vol. 79, 1986, pp. 239-248.

2.   Ledbetter, H. M., Datta, S. K., and Kyono, T., "Elastic Constants of a Graphite-Magnesium Composite," to appear in the Journal of Applied Physics.

3.   Mindlin, R. D., "Waves and Vibrations in Isotropic Elastic Plates," Proceedings of the First Symposium on Naval Structural Mechanics, Pergamon Press, 1960, pp. 114-128.

4.   Newman, E. G. and Mindlin, R. D., "Vibrations of a Monoclinic Crystal Plate," Journal of the Acoustical Society of America, Vol. 29, 1957, pp. 1206-1218.

5.   Kaul, R. K. and Mindlin, R. D., "Vibrations of an Infinite, Monoclinik Crystal Plate at High Frequencies and Long Wavelengths," Journal of the Acoustical Society of America, Vol. 34, 1962, pp. 1895-1901.

6.   Mindlin, R. D. and Medick, M. A., "Extensional Vibrations of Elastic Plates", Journal of Applied Mechanics, Vol. 26, 1959, pp. 561-569.

7.   Dong, S. B. and Pauley, K. E., "Plane Waves in Anisotropic Plates," ASCE Journal of Engineering Mechanics, Vol. 104, 1977, pp. 801-817.

8.   Dong, S. B. and Huang, K. H., "Edge Vibrations in Laminated Composite Plates,"

Jornal of Applied Mechanics, Vol. 52, 1985, pp. 433-438.

9.   Datta, S. K., Shah, A. H., Al-Nassar, Y., and Bratton, R. L.,   "Elastic Wave Dispersion in Laminated Composite Plate,"   in Review of Program in Quantitation Nondestructive Evaluatioon, edited by D. O. Thompson and D. E. Chimenti, Plenum Press, 1988, pp. 987-994.

10. Datta, S. K., Shah, A. H., Bratton, R. L., and Chakraborty, T., "Wave Propagation in Laminated Composite Plates," Journal of the Acoustical Society of America, Vol. 83, 1988, pp. 2020-2026.

11.   Chimenti, D. E. and Nayfeh, A. H., "Leaky Lamb Waves in Fibrous Composite Plates," Journal of Applied Physics, Vol. 58, 1985, pp. 4531-4538.

12.   Nayfeh, A. H. and Chimenti, D. E., "Program of Guided Waves in Fluid-coupled Plates of Fiber-Reinforced Composite," Journal of the Acoustical Society of America, Vol. 83, 1988, pp. 1736-1743.

13.   Buchwald, V. T., "Rayleigh Waves in Transversely Isotropic Media," Quarterly Journal of Mechanics and Applied Mathematics, Vol. 14, 1961, pp. 293-317.

Interface Development in SiC Fiber-Reinforced Titanium Aluminide Matrix Composites

S. M. JENG AND J.-M. YANG

ABSTRACT

The interface development in SiC (SCS-6) fiber-reinforced two titanium aluminides (Ti-24Al-11Nb, Ti-25Al-10Nb-3V-1Mo) matrix composites was studied. The microstructure and elemental distributions of the reaction zone were characterized using electron microscopy and electron probe microanalysis. It was found that the matrix alloy composition will affect the microstructure and the distribution of the reaction products, as well as the growth kinetics of the reaction zones.

INTRODUCTION

The titanium aluminides, Ti_3Al, TiAl, are promising materials for high temperature structural applications [1]. They have lower density, higher modulus and elevated-temperature strength, better creep resistance, and better resistance to high-temperature oxidation than conventional titanium alloys. However, the wide-spread applications of the intermetallic alloys have been limited due to their low ductility and brittle intergranular fracture at ambient temperature. Nevertheless, significant improvement in the ductility and strength of titanium aluminides has been achieved recently by various meta- llurgical techniques such as microalloying, grain refinement, and rapid solidi- fication. For example, the addition of β-stabilizing elements such as Nb, Mo, and V can substantially increase the strength and ductility of Ti_3Al [2,3]. These elements will stabilize the high temperature body-centered cubic form of titanium at low temperatures resulting in alloys with aluminide + beta struc- tures which have adequate low temperature ductility.

S. M. Jeng, Graduate Research Assistant, J.-M. Yang, Assistant Professor, Department of Materials Science and Engineering, University of California, Los Angeles, CA 90024.

The incorporation of low density, high strength and stiffness continuous fibers can significantly enhance the mechanical properties of titanium alumi-nide. It has been demonstrated that the SiC fiber-reinforced titanium aluminide composites have much better specific strength than that of nickel-base super-alloys [4]. However, titanium aluminides are thermodynamically incompatible with most of the reinforcing fibers [5-8]. The formation of brittle reaction pro-ducts at the interface can seriously degrade the mechanical properties. As a result, one of the most challenging tasks in developing these novel materials is to prevent the damaging fiber/matrix interfacial reactions during consolidation and service.

The objective of this paper was to study the effect of matrix alloy compo-sitions on the interface developemnt and stability of SiC fiber-reinforced titanium aluminide composites.

EXPERIMENTAL PROCEDURE

The composites used in this study were Ti-24 Al-11 Nb and Ti-25 Al-10 Nb-3 V-1Mo reinforced with silicon carbide (SCS-6) fibers. The unidirectional composite panels were consolidated by the vacuum diffusion bonding technique. The microstructure of the interface was characterized using optical and scan-ning electron microscopy. The chemical compositions of the reaction zones were analyzed using electron probe for microanalysis (EPMA). In order to study the reaction kinetics, the composite specimens were encapsulated in a quartz tube at a vacuum of 10^{-3} torr, and thermally exposed at 700 - 1000 °C for 2 to 200 hours. After thermal exposure, the specimens were prepared for meta-llographic examination. The thickness of the reaction zone was measured through high magnification SEM micrographs.

RESULTS AND DISCUSSION

The scanning electron micrographs of the as-consolidated SCS-6 fiber in Ti-24-11 and Ti-25-10-3-1 are shown in Fig. 1. It is evident that both types of matrix alloys contain mixtures of α_2 and β phase. However, more amount of β-phase was found in Ti-25-10-3-1 alloy due to the addition of more β-stabilizing elements. In these figures, it can be seen that a reaction zone between fiber and matrix has formed during consolidation. The thickness of the reaction zone is 1.62 and 0.93 mm for the as-fabricated SCS-6/Ti-24-11 and SCS-6/Ti-25-10-3-1, respectively. Furthermore, a thin layer of β-depleted

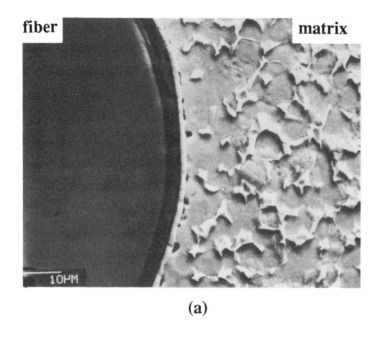

fiber   **matrix**

10µM

(a)

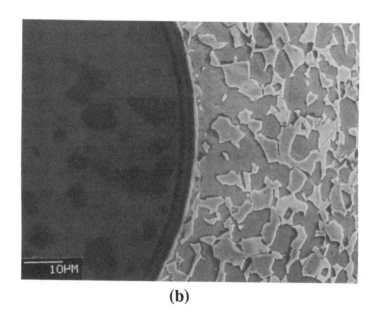

10µM

(b)

Fig. 1 The scanning electron micrographs of the as-fabricated SCS-6 fiber in
(a) Ti-24Al-11Nb, and (b) Ti-25Al-10Nb-3V-1Mo.

region immediately adjacent to the reaction zone was observed in SCS-6/ Ti-24-11 composite. The formation of the β–depleted zone might be due to Nb and Ti being consumed proportionally faster than the Al (an α stabilizer) in the reaction with carbon [8]. With a higher amount of β-stabilizing elements in the Ti-25-10-3-1 alloy, β-depleted zone was not observed at the interface. The exact mechanism of formation of the β-depleted zone is still under investigation.

The detailed chemical compositions across the reaction zone of each composite were analyzed by EPMA. The backscattered electron image and elemental x-ray line scan across the reaction zone for both types of composites after annealing at 950 °C (1742 °F) for 8 hours are shown in Figs. 2 to 3, respectively. The backscattered electron image revealed that the reaction zone was composed of a multilayer reaction products. X-ray line scan clearly indicates that extensive interdiffusion of the atomic species in the fiber and matrix occurred. All the alloy elements in Ti-24-11 matrix diffused into the interface layer and participate in the reaction as shown in Fig. 2(b). Si and C also diffused out to react with matrix alloy. The reaction products have been identified as composed of $(Ti, Nb)C_{1-x}$, $(Ti, Nb,Al)_5Si_3$ adjacent to the fiber; $(Ti, Nb)_3AlC$ and $(Ti, Nb,Al)_5Si_3$ adjacent to the matrix [8]. In the SCS-6/Ti-25-10 -3-1 system, the distributions of Ti, Nb, Al, and Si are similar to those in SCS-6/ Ti-24-11 composite. However, some minor amount of V was also detected in the reaction zone. The detailed microstructure of the reaction products has not been identified yet.

The thickness of the reaction zone developed progressively as the annealing temperatures and time increased. Fig. 4(a) shows the growth of the reaction zone as a function of square root of time at 950 °C. It is obvious the growth rate in SCS-6/Ti-25-10-3-1 composite is faster than that in SCS-6/ Ti-24-11. The consumption of the C-rich layer on the SCS-6 fiber as a function of time at 950 °C is given in Fig. 4(b). The C-rich layer on the SCS-6 fiber in Ti-25-10-3-1 was also consumed faster than that in Ti-24-11 matrix. This might be due to the presence of more β-phase in Ti-25-10-3-1 alloy. Previous study showed that carbon diffused at a higher rate in β phase than in α phase, hence the fiber surface layer would be expected to be consumed at a faster arte wherever it is in contact with beta phase [9]. Therefore, SCS-6 fiber-reinforced titanium alloy composite exhibited increased reaction zone formation with increased volume fraction of β phase. Initially, the reaction zone was formed by the reaction of matrix alloy and the C-rich outer layer of the fiber. However, when the C-rich

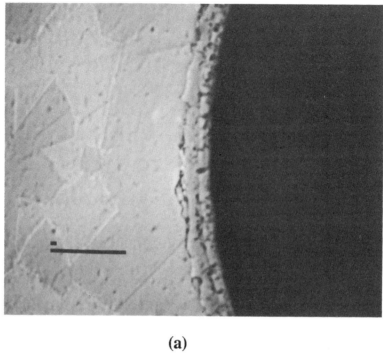

(a)

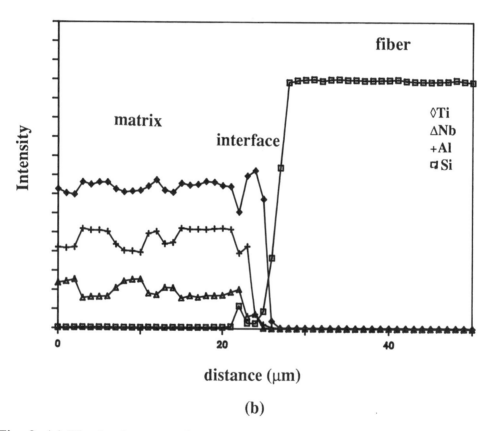

(b)

Fig. 2 (a) The backscattered electron image, and (b) X-ray line scan across
the reaction zone for SCS-6/Ti-24Al-11Nb composite after thermal
exposure at 950 °C for 8 hours.

46

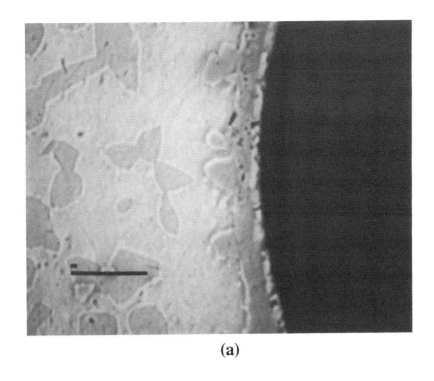

(a)

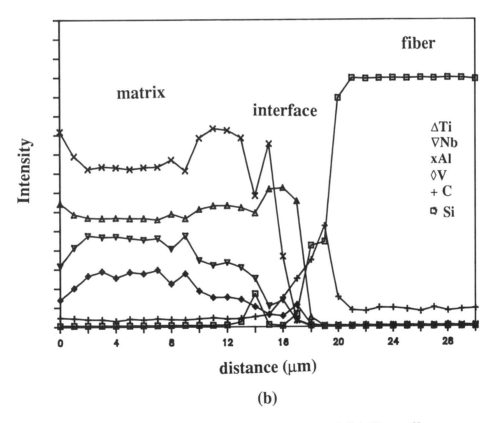

(b)

Fig. 3 (a) The backscattered electron image, and (b) X-ray line scan across
the reaction zone for SCS-6/Ti-25Al-10Nb-3V-1Mo composite after
thermal exposure at 950 °C for 8 hours.

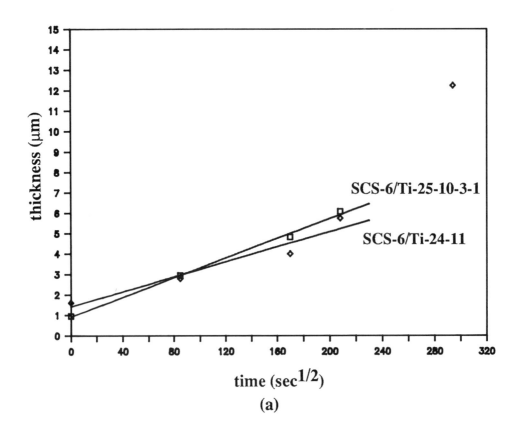

(a)

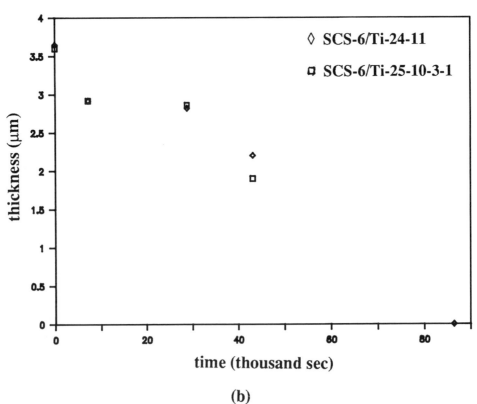

(b)

Fig. 4 (a) The growth of the interfacial reaction zone as a function of square root of time at 950 °C, and (b) the comsumption rate of C-rich layer as a function of time at 950 °C for SCS-6 fiber-reinforced Ti-24Al-11Nb and Ti-25Al-10Nb-3V-1Mo composites.

layer was consumed, the thickness of the reaction zone increased rapidly for both types of composites as indicated in Fig. 4(a).

SUMMARY

(1) SCS-6 fiber reacted extensively with Ti-24Al-11Nb and Ti-25Al-10Nb-3V -1Mo to form complex carbides and silicides at the interface during consolidation and high temperature exposure.

(2) The matrix alloy composition will affect the microstructure and the distribution of the reaction products, as well as the growth kinetics of the reaction zones.

REFERENCES

1. Lipsitt, H. A., "Titanium Aluminides--An Overview", High-Temperature Ordered Intermetallic Alloys, edited by C. C. Koch et. al., Materials Research Society, Pittsburgh, 1985, pp. 351-364.

2. Martin, P. L., Lipsitt, H. A., Nuhfer, N. T., and Williams, J. C., "The Effect of Alloying on the Microstructure and Properties of Ti_3Al and TiAl", Proceedings of the 4th International Conference on Titanium, Japan, 1980, pp. 1245.

3. Khataee, A., Flower, H. M., and West, D. R. F., "New Titanium-Aluminum-X Alloys for Aerospace Applications", Journal of Materials Engineering, Vol. 10, 1988, pp. 37-44.

4. Brindley, P. K., "SiC Reinforced Aluminide Composites", High-Temperature Ordered Intermetallic Alloys II, edited by N. S. Stoloff et. al., Materials Research Society, Pittsburgh, 1987, pp. 419-426.

5. Yang, J.-M., and Jeng, S. M., "Interfaces in SiC Fiber Reinforced Titanium Aluminide Matrix Composites", Metallurgical Transactions, 1989, in press.

6. Jeng, S. M., Shih, C. J., W. Kai, and Yang, J.-M., "Interface Reaction Studies of B_4C/B and SiC/B Fiber-Reinforced Ti_3Al Matrix Composites", Materials Science and Engineering, 1989, in press.

7. Jeng, S. M., and Yang, J.-M., "Kinetics of Interfacial Reactions in Fiber-Reinforced Ti_3Al + Nb Matrix Composites", Proceeding of the 7th International Conference on Composite Materials, Beijing, 1989, in press.

8. Baumann, S. F., Brindley, P. K., and Smith, S. D., "Reaction Zone Microstructure in a Ti$_3$Al + Nb/SiC Composite", <u>Metallurgical Transactions</u>, 1989, in press.

9. Rhodes, C. G., and Spurling, R. A., "Fiber-Matrix Reaction Zone Growth Kinetics in SiC-Reinforced Ti-6Al-4V as Studied by Transmission Electron Microscopy", <u>Recent Advances in Composites in the United States and Japan</u>, ASTM STP 864, edited by J. R. Vinson, and M. Taya, American Society for Testing Materials, Philadelphia, 1985, pp. 585-599.

Nondestructive Characterization of Metal-Matrix Composite Microstructures

R. E. SHANNON, P. K. LIAW AND W. G. CLARK, JR.

ABSTRACT

This paper presents the results of laboratory studies directed toward developing nondestructive evaluation (NDE) techniques for the serviceability assessment of metal matrix composite materials. Ultrasonic, eddy current and electrical resistivity techniques are used to measure composite microstructural features that control material performance properties. A linear superposition model is used to develop a correlation between NDE signatures and microstructural characteristics of the composites. Combined mode NDE techniques are shown to provide a rational, real-time approach to the measurement and qualification of the complex microstructural characteristics of metal-matrix composites. The results of this program can be used to establish practical guidelines for the in-process and in-service inspection and qualification of composite materials.

INTRODUCTION

The use of "state-of-the-art" composites is currently limited by the absence of high reliability material qualification methods. Specifically, metal matrix composite materials provide a significant challenge for conventional nondestructive evaluation (NDE) techniques. By their very nature, composites essentially consist of controlled "discontinuity" structures. Consequently, the popular penetrating energy volumetric NDE methods, such as ultrasonics, radiography, and eddy current testing often yield poor signal-to-noise ratio data that can mask defects. As the differences in physical properties between components (matrix and reinforcement) increase, the relevant versus non-relevant aspects of inspection data interpretation become more complex.

The value of advanced data processing or any other aspect of NDE cannot be assessed without a detailed consideration of the acceptance/rejection criteria for the material. This consideration highlights another important limitation in the use of metal matrix composites - understanding of the failure mechanisms. Like any structural material, the predominant fracture control mechanisms in a composite can vary with the specific loading conditions and service environment. However, unlike more

Westinghouse Research & Development Center, 1310 Beulah Road, Pittsburgh, Pennsylvania 15235.

conventional monolithic materials, relatively little mechanistic work has been done and the complexities of dissimilar interacting components prevent accurate prediction of fracture performance. In addition, virtually no service experience data base exists and even systematic evaluation of anything but the most common tensile properties are rare. Consequently, guidelines for the acceptance and qualification of MMC components do not exist.

This situation is not adequate for the efficient use and continued growth of the MMC material option. The objectives of our present research is to develop NDE techniques with a sufficient data base for use in the assessment of MMC serviceability. We will show that some common NDE methods have the ability to distinguish microstructural variations in Al/SiC composite materials. This is an important step toward the objective of serviceability assessment because the microstructure controls material properties. The results presented in this paper will therefore serve as baseline characteristics for subsequent mechanical performance testing of these materials.

MATERIALS AND EXPERIMENTAL PROCEDURE

Silicon-carbide particle reinforced, aluminum (Al/SiCp) alloy composite materials, purchased from DWA Composite Specialties, Inc., were used in the present study. They were fabricated by the powder metallurgy technique and supplied in the as-extruded condition. Detailed information of the sample material are given in Table 1.

The samples were nondestructively characterized by measuring ultrasonic velocity, direct-current (D.C.) electrical resistivity and eddy current response from selected locations on the as-received extrusions. Both longitudinal and shear velocities were determined using the pulse-echo ultrasonic method. Accurate time-base readings were made to the nearest 0.01 μsec with a calibrated digital oscilloscope.

An AT&T, Model 100 Microhmeter was used to obtain D.C. resistivity measurements. This portable instrument uses the four-point contact probe technique and provides direct readings in microhm-centimeters (μohm-cm).

Eddy current measurements were made using techniques previously developed on laboratory structural models of the composite materials [1]. Three test frequencies, 1 kHz, 5 kHz and 700 kHz, were used to energize the materials. These frequencies were determined as likely to provide the optimum sensitivity to the characterization of the Al/SiCp structures at varying penetration depths. The eddy current measurements were made using a standard field portable test instrument and surface contact probes designed for optimum response at the test frequencies.

Microstructural characterization of the composite materials were performed by machining metallographic specimens from areas of the samples representative of the NDE test locations. The specimens were prepared for microstructural examination in a scanning electron microscope (SEM). The SEM investigations included identification of SiCp, intermetallic compounds and porosity. The loading and distribution of SiCp were determined from secondary-electron-image photographs. The presence and volume fraction content of intermetallic compounds and porosity were determined from back-scattered-electron-image photographs.

TEST RESULTS

Figures 1(a) and 1(b) show plots of the measured longitudinal and shear velocities versus SiCp loading for each of the three composite alloy systems. The velocities measured in each of the unloaded (0 v/o SiCp) samples are found to agree well with the range of velocities for wrought aluminum alloys [2,3]. For example, aluminum alloy longitudinal velocities typically range between 6.25×10^5 and 6.35×10^5 cm/sec and shear velocities typically range between 3.08×10^5 and 3.13×10^5 cm/sec. The influence of SiCp loading is seen to be a nearly linear increase in measured velocity with increasing loading, although some slight deviations are evident.

Figures 2(a) and 2(b) show plots of the average measured eddy current response versus SiCp loading at the highest and lowest test frequencies investigated. As seen in the figures, reasonably good correlations exist between eddy current response and SiCp loading for the 2124 Al and 7091 Al composites. They show that increasing SiCp loading generally increases the eddy current response. Some variations in eddy current responses were observed, however, that are not directly related to SiCp loading. Although the intermediate frequency results are not shown, the variations were noted as being consistently greater with increasing frequency. Of particular interest are the two separate 2124 Al composites that were similarly loaded to 25 v/o SiCp (billets PE-2404 and 2229 from Table 1). The eddy current readings from PE-2229 are consistently higher than from PE-2404. It is also noted that the eddy current response of the 30 v/o SiCp loaded 2124 Al composite was lower than PE-2229 (25 v/o SiCp).

The eddy current responses are observed to be significantly different for the 6061 Al group of composite samples. The eddy current responses were lower overall than for the other two alloy groups, although the eddy current response did not uniformly increase with SiCp loading. For example, the 20 v/o SiCp loaded 6061 Al composite (PE-2047) shows a much greater reading than both the 25 v/o and 30 v/o samples in the same group.

The results of the four-point D.C. resistivity measurements are shown in Figure 3. The absolute resistivity values of the unloaded (0 v/o) samples compare well with the available handbook values of 5.65, 4.31 and 5.5 μohm-cm for the three alloys in their respective heat treated conditions [4]. The observed trends in the resistivity measurements are similar to the eddy current responses. The lower overall eddy current response for the 6061 Al composite group is related to the lower resistivity of the matrix alloy. Deviations from a uniform increase based on added SiCp loading are also seen, and are similar to those found in the eddy current responses. For example, the resistivity reading from PE-2229 is higher than PE-2404, although both are loaded with 25 v/o SiCp, and also higher than the 30 v/o SiCp sample. Most noticeably, the the resistivity measured for the 20 v/o SiCp loaded 6061 Al composite was 3 μohm-cm higher than might be expected based on the trend observed from the other samples.

Figures 4(a) and 4(b) exhibit typical SEM photographs used for microstructural characterization of the composites. These examples show the SEM secondary-electron and back-scattered-electron images taken from the 30 v/o SiCp loaded 6061 Al composite sample. The secondary-electron image clearly demonstrates the morphology and distribution of the SiCp.

The companion back-scattered-electron image is seen to be particularly useful for identifying both intermetallic compounds and porosity.

The volume percentages of SiCp, intermetallic compounds and porosity were quantitatively measured by the point counting method [5]. The SiCp loadings were measured, and confirmed as within the nominal values provided by the supplier. All of the composites, including the unloaded (0 v/o SiCp) samples, are found to contain intermetallic compounds. These intermetallic compounds are exhibited as white on the back-scattered-electron image [Figure 4(b)], and show as relatively fuzzy shapes on the secondary-electron images [Figure 4(a)]. An EDS analysis was performed using the SEM and the intermetallic compounds were identified as containing various combinations of elements, including Al, Si, Mn, Fe, Cu, Cr, Mg, Co, Zn, Ti and Zr, depending on the base alloy. The quantitative results of the average intermetallic compound content for each sample are shown in Table 1, with a large variation seen within each composite group. In particular, the 20 v/o SiCp loaded 6061 Al sample exhibits an unusually high intermetallic compound content, which correlates with the high eddy current response and resistivity values in Figures 2 and 3. It has a 15.5 v/o intermetallic compound content compared to 5.2, 2.9 and 1.2 v/o in the 0, 25 and 30 v/o SiCp loaded 6061 Al composites. Also, the greater volume percent of intermetallic compounds in PE-2229 correlates with higher eddy current response and resistivity values when compared with PE-2404.

The presence of porosity is found in each of the 30 v/o SiCp loaded composites. The porosity content varies from 1.4 to 4.2 v/o in the 30 v/o SiCp samples. The only other sample found with porosity is the 10 v/o SiCp loaded 7091 Al, measured at 0.5 v/o. The porosity is generally found to be present at interfaces between the SiCp and the matrix. Note that the porosity is exhibited as black on the back-scattered-electron image [Figure 4(b)].

DISCUSSION

Although the ultrasonic velocities show a relatively uniform increase with added SiCp loading, it is clear that both the eddy current and D.C. resistivity readings are sensitive to other features of the microstructures. Earlier modeling studies [1] showed that the presence of porosity should cause a slight decrease in ultrasonic velocity as a simple function of porosity content. Similar relationships, however, were not developed for intermetallic compounds since their inherent velocities relative to aluminum are not easily known.

The modeling studies showed that the presence of porosity can be expected to cause a measurable deviation in eddy current response. The effect is mainly caused by the contribution of porosity to increased electrical resistivity. Also, the increased eddy current response with added SiCp loading may be attributed to the resistivity of silicon carbide being at least two orders of magnitude greater than aluminum [6-8]. The D.C. resistivity results are straightforward in this respect. Although the effects of intermetallic compounds were not modeled, it is likely that these intermetallic compounds have electrical resistivity values exceeding aluminum. Also, any slight magnetic permeability variations arising from Fe-bearing compounds will greatly affect eddy current response. Porosity and intermetallic compounds are undesirable constituents in a composite since they are known to degrade material strength [9]. The presence and content level of both intermetallic compounds and porosity, as well as SiCp loading, are, therefore, used to correlate the NDE and microstructural characterization results.

A linear superposition model is used to correlate the composite samples NDE signatures to these microstructural features. Equation (1) describes the linear superposition model,

$$AX + BY + CZ + DW = N \tag{1}$$

where X, Y and Z are SiCp, intermetallic and porosity contents, respectively. W is aluminum v/o, where W = 100 - (X + Y + Z). A, B, C and D are coefficients developed for SiCp, intermetallics, porosity and aluminum, respectively; N is the measurement magnitude for each NDE technique taken individually. The results of the present study are used to develop a solution for the four correlation coefficients A, B, C and D. Three independent linear superposition equations can then be obtained to solve for the three independent unknowns (X, Y and Z) using the three sets of NDE results: ultrasonic velocities, eddy current responses and resistivity values. Thus, in principle, three microstructural features (SiCp, intermetallic compounds and porosity) can be predicted nondestructively.

The value, D, for the ultrasonic equation is estimated by a simple extrapolation to 0 v/o SiCp of the nearly linear relationship between velocity and SiCp loading. Separate plots (not shown) for eddy current response and resistivity were found to be nearly linear versus total constituent loading (SiCp loading + intermetallic content + porosity content). The values, D, for eddy current and resistivity equations are then estimated by extrapolation to 0 v/o total loading.

For the ultrasonic equation, the nearly linear correlation to SiCp v/o allows simplification of Equation (1) by setting coefficients B and C equal to zero. Data from the 2124 Al composite group are used to provide an ultrasonic velocity value for A. As discussed, the eddy current and resistivity signatures are not only a function of SiCp loading, but also significantly affected by intermetallic compound and porosity content. In each composite group studied, there are at least four composite materials with various volume percentages of SiCp, intermetallic compounds and porosity. Thus, using the NDE results from one of the composite systems, 2124 Al, the values of the correlation coefficients A, B and C are estimated.

The results of ultrasonic longitudinal and shear velocity analyses are used to predict the SiCp loading. Figures 5(a) and 5(b) compare the predicted SiCp loading versus the loading measured from the microstructural characterization. The predicted SiCp loadings are found to be in good agreement with the measured values for all three composite matrix alloy groups, even when using one set of coefficients for Equation (1). After the SiCp loading is predicted, the content of the other two independent constituents, intermetallics and porosity, are estimated. In principle, two linear superposition equations based on the remaining two NDE signatures, eddy currents and resistivity, can be used to predict these constituents. Attempts to reach a solution based on the results achieved in this study do no allow independent prediction of intermetallics and porosity. It is believed that the eddy current and electrical resistivity measurements are not sufficiently independent measurement variables for this purpose.

Alternatively, the present data is useful in predicting the total content of intermetallics plus porosity. Figure 6 shows the results of using eddy current and resistivity measurements and the linear superposition model to predict these constituents. The plot shows that the predicted total content of intermetallics plus porosity is consistent with the total content measured by microstructural characterization. It should be noted that the prediction of constituent loadings is based on the coefficients of SiCp, intermetallic compounds and porosity, which are empirically determined using the NDE signature from only one composite matrix alloy group. The predicted values are found to be in good agreement with the measured values regardless of the composite group investigated. This trend suggests that the coefficients, A, B and C, may be intrinsic material constants which are insensitive to the matrix alloy. Further analysis also shows consistency between coefficients based on known physical properties. For example, using ultrasonic velocities, coefficient A (SiCp) is found to be approximately 1.5 to 2.0 times greater than coefficient D (aluminum). This behavior seems in line with the experimental findings that the ultrasonic velocity of SiC is about 1.7 times greater than that of aluminum alloy [9,10].

SUMMARY

A simple linear superposition equation has been developed to model the correlation between NDE signature and Al/SiCp composite microstructural features. The measurement of ultrasonic velocities can be used to nondestructively predict the SiCp volume percentage in aluminum matrix composite materials. Eddy current and electrical resistivity measurements are found to be very sensitive to the presence of intermetallic compounds in Al/SiCp composites. In the present composite materials, the combined total volume percentage of intermetallic compounds and porosity can be predicted nondestructively by the eddy current and resistivity measurements.

REFERENCES

1. W. G. Clark, Jr. and J. N. Iyer, "Structure Modeling and the NDE of Metal Matrix Composites", Materials Evaluation (in press).

2. J. Krautkramer and H. Krautkramer, Ultrasonic Testing of Materials, Second Edition, Springer-Verlag, Berlin Heidelberg, New York, 1977.

3. Metals Handbook, Volume 11, "Nondestructive Inspection and Quality Control", Eighth Edition, ASM, Metals Park, OH, 1975.

4. Metals Handbook, Volume 1, "Properties and Selection of Metals", Eighth Edition, ASM, Metals Park, OH, 1975.

5. J. E. Hilliard and J. W. Cahn, Transactions of the Metallurgical Society of AIME, vol. 221, 1961, p. 334.

6. W. E. Nelson, Stanford Research Institute, "Beta-Silicon Carbide and its Potential for Devices", FR No. 21, Contract No. 87235, Dec. 31, 1963.

7. <u>Proceedings of the International Conference on Silicon Carbide</u>, Materials Research Bulletin, H. K. Henisch and R. Roy, eds., Pergamon Press, 1968.

8. <u>Silicon Carbide-1973</u>, R. C. Marshall, J. W. Faust, Jr., and C. E. Ryan, eds., University of South Carolina Press, Columbia, SC, 1974.

9. G. Mott and P. K. Liaw, Metallurgical Transactions A, vol. 19A, 1988, p. 2233.

10. E. R. Generazio, D. J. Roth and G.Y. Baaklini, Materials Evaluation, 46, 1988, p. 1338.

Table 1 - Identification of Al/SiCp Sample Material
and Measured Microstructural Features

Matrix Alloy	Billet Number	SiCp Loading	Extrusion Ratio	Intermetallic Content	Porosity Content
2124-T4	PE-2600	0%	20:1	7.4%	0%
2124-T4	PE-2404	25%	11:1	4.4%	0%
2124-T4	PE-2229	25%	11:1	10.0%	0%
2124-T4	PE-2488	30%	20:1	6.7%	1.4%
6061-T6	PE-2045	0%	11:1	5.2%	0%
6061-T6	PE-2047	20%	10:1	15.5%	0%
6061-T6	PE-2099	25%	39:1	2.9%	0%
6061-T6	PE-2731	30%	20:1	1.2%	2.6%
7091-T6	PE-2730	0%	20:1	6.9%	0%
7091-T6	PE-2711	10%	25:1	6.9%	0.5%
7091-T6	PE-2712	20%	25:1	4.4%	0%
7091-T6	PE-2713	30%	20:1	3.2%	4.2%
7091-T6	PE-2665	30%	11:1	6.9%	1.6%

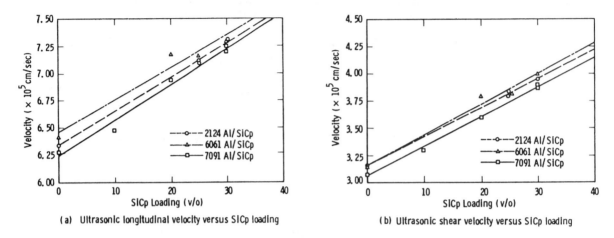

(a) Ultrasonic longitudinal velocity versus SiCp loading

(b) Ultrasonic shear velocity versus SiCp loading

Fig. 1. The influence of SiCp loading on the measured ultrasonic velocity in aluminum matrix composites.

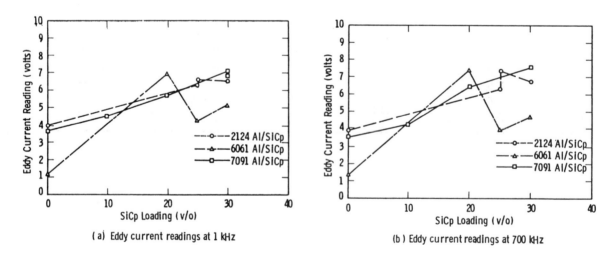

(a) Eddy current readings at 1 kHz

(b) Eddy current readings at 700 kHz

Fig. 2. Eddy current response of aluminum matrix composites at various SiCp loadings.

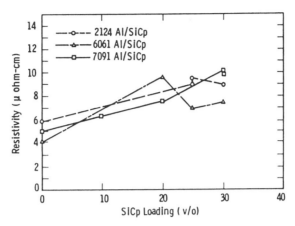

Fig. 3. Four-point D.C. surface resistivity response of aluminum matrix composites at various SiCp loadings.

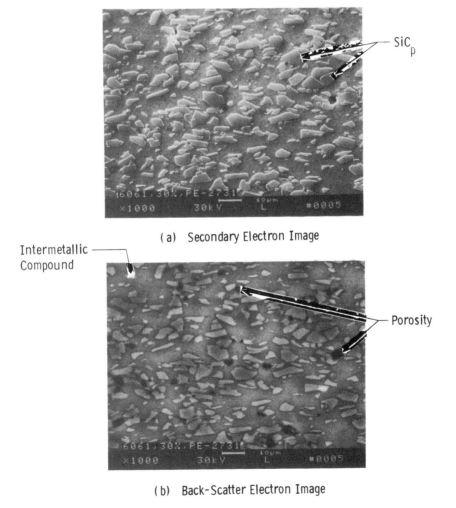

(a) Secondary Electron Image

(b) Back-Scatter Electron Image

Fig. 4. Microstructure of 6061 Al/30 v/o SiCp composite (PE-2731).

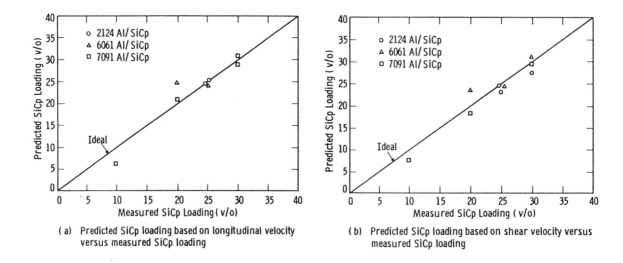

(a) Predicted SiCp loading based on longitudinal velocity versus measured SiCp loading

(b) Predicted SiCp loading based on shear velocity versus measured SiCp loading

Fig. 5.   Prediction of SiCp loading using ultrasonic velocity measurements.

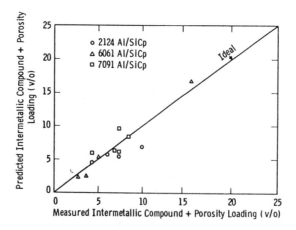

Fig. 6.   Prediction of intermetallic and porosity content using eddy current response and resistivity measurements.

Microstructure and Mechanical Properties of High Temperature Aluminum Composites

WM. D. POLLOCK
University of Virginia
NASA Langley Research Center
Hampton, Virginia 23665
MS 188-E

F. E. WAWNER
University of Virginia
Materials Science
Thornton Hall
Charlottesville, Virginia 22901

Introduction

To increase performance and efficiency, future aerospace vehicles will have components, both engine and structural, which will be exposed to elevated temperatures. Engines will burn hotter to increase efficiency, and speeds will be great enough to heat the leading edges of wings to 1000°C plus. Temperatures considered in this program range up to 482°C. Titanium alloys have the required strength at these temperatures, but there is a weight and cost penalty.

The light weight metals, such as aluminum, currently available don't have the required strength at these temperatures. One of the methods to increase the strength of light metals is to make a composite by adding fiber reinforcement. The goal of this project was to develop a high strength, light weight composite that had good elevated temperature strength.

Composite reinforcement is attractive as it can increase the specific strength and stiffness of light metals over a wide temperature range without significantly increasing weight, hence this approach appears to offer the highest probability to attain better high temperature properties for an aluminum material.

The first step was to understand what mechanisms limit the elevated temperature strength of composites. Once these mechanisms were defined and understood the microstructure could be altered through alloying and processing to improve upon temperature limitations. The transfer of load to the reinforcement is what limits the elevated temperature strength.

The most important variables in transferring load to the reinforcement are the aspect ratio of the reinforcement, and the shear strength of the matrix and/or the interface. Typically the shear strength of alloys decreases as temperature increases. This fact requires the fiber aspect ratio and/or fiber strength to be increased to achieve high temperature composite strength.

The shear strength of aluminum alloys can be increased by adding fine intermetallic dispersoids through rapid solidification [1]. However, rapidly solidified alloys are extremely sensitive to processing since exposure to high temperature will coarsen the dispersoids very rapidly minimizing their influence.

Al-8Fe-2Mo is a dispersion hardened alloy produced by rapid solidification. It was used in this study since it has encouraging high temperature strength. Al_2O_3 (Saffil) fibers were used as the reinforcement.

Experimental
The powders were made by ALCOA using inert gas atomization. The as received (AR) aluminum powder is of irregular size and morphology as shown in Figure 1.

The Saffil, alumina fibers from ICI (Mond Division, Cheshire England) have an initial average aspect ratio of approximately 20:1. The surface of the fibers is uniform, but some of the fibers have some curvature and their diameter varies, see Figure 2. The Saffil has a UTS of 2000 MPa [e].

Composites are usually processed at high temperatures to facilitate a reaction between the fiber and matrix to improve the strength of the interface. Temperatures required for this would coarsen the dispersoids in the matrix and decrease the shear strength. Some of the fibers were coated with cobalt for evaluation of low temperature bonding capability. This cobalt coating can act as an intermediate layer, reacting with the matrix at lower temperatures to improve the strength of the interface with minimal coarsening of the dispersoids. The reinforcing fibers were coated by Boeing Aerospace using their proprietary sol gel process [3].

The powder was mixed with the reinforcement in house using ethyl alcohol as the mixing medium in a household blender. Each charge was then sent to INCO Alloys (Huntington, WVa) for canning and extrusion. The powders were vacuum degassed in the can at 482°C for several hours to remove water and hydrated oxides [4,5,6], and then hot compacted with a pressure of 139 MPa. The samples were then extruded at a ratio of 16:1, at 482°C.

Tensile bars were machined from the extruded rods so the tensile axis coincided with the extrusion axis and therefore the axis of the fibers. The samples were tested in three conditions: as received (AR), heat treated for 10 hours at 400°C (HT4), and for 120 hours at 500°C (HT5). The heat treatments were done in air with a furnace cool. The elevated temperature tensile tests were performed in a resistance heater and held at temperature for one half hour before testing to insure thermal equilibrium. The thermocouple used to control the temperature was attached to the sample and shielded from the heater with high temperature cloth.

Results and Discussion
Samples extruded from the Al-8Fe-2Mo/20 $^v/_o$ Al_2O_3 display the microstructure shown in Figure 3. It can be seen that consolidation is less than perfect with some voids, fiber clumping, and aluminum rich pockets. In addition the length of the fibers has been considerably reduced due to the extrusion. Original aspect ratios of approximately 20 are in the range of 1 to 6 for the extruded composite with a mean value of 2.4.

It was very difficult to extrude this material due to the low temperatures used to prevent coarsening of the matrix; however, enough was obtained to get test specimens for room temperature and three elevated temperature tests (400, 450, 500°C).

During the background study for this project a search of literature yielded the data shown in Figure 4. In this figure 5 different matrices, 2 different reinforcements, 4 fabrication conditions, and 2 orientations are represented. Strength values are widely different at room temperature but at 350°C and above all the strengths converge to a narrow and continuously converging band at approximately 60 MPa. Hence it appears that none of the factors mentioned above influence the strength when tested at temperatures above 350°C.

Other investigations [7] have demonstrated that fracture at temperatures above 350°C is controlled by the nature of the matrix and in order to get higher tensile properties for an aluminum material above 350°C a matrix with stable dispersoids to inhibit slip and develop higher shear strength is necessary. Therefore using the Al-8Fe-2Mo rapidly solidified alloy as the matrix should be a beneficial approach.

The strength of each system in each condition is presented for each test temperature in Figure 5. The unreinforced matrix in the AR condition is stronger than the heat treated (HT5) sample at room temperature. The coarsening of the dispersoids during heat treating reduces the room temperature shear strength. As the test temperature increases the strength of the matrix decreases. Deformation at elevated temperature is less dependent on the size and spacing of dispersoids and therefore the HT5 sample is stronger than the AR sample.

The AR Saffil reinforced composite room temperature strength is comparable to the AR matrix. The short aspect ratio fibers combined with a weak interface prevent much improvement in the strength. After heat treating, the room temperature strength is further reduced due to coarsening of the matrix. The HT4 has greater room temperature strength since the thermal exposure is less severe than the HT5.

As the test temperature is increased the strength decreases. The heat treated samples, specifically the HT4, are stronger than the AR sample at elevated temperature. This is again due to the change in deformation at elevated temperature and the relationship to the distribution of dispersoids in the matrix.

The Co/Saffil composite is the only system where the room temperature strength improves after heat treating. In the AR condition the Co/Saffil composite was weaker than the matrix alone, but during heat treating the cobalt reacts with the matrix, improving the strength of the interface. Since the thermal exposure of the HT4 treatment is less severe, the dispersions coarsened less, and the HT4 sample is stronger than the HT5 sample at room temperature.

The strongest system at each temperature is one containing Co/Saffil. At 500°C the heat treated Co/Saffil samples are the strongest system studied (.51 MPa). The HT4 system has the best strength overall. Again, it combines the stronger interface of the heat treated Co/Saffil system with the finer dispersoid distribution of the HT4 treatment.

Fracture surfaces were studied by a Jeol 35 scanning electron microscope. There are several different types of defects on the fracture surfaces. They exist in every system and at every temperature. Inclusions containing Yttrium and Chromium exist in the atomized powder and are seen as defects as in Figure 6. ALCOA produces a rapidly solidified aluminum alloy containing Y and Cr, so the source of the defect must be a contaminated atomizer. Other defects come from shot in the Saffil (Figure 7) and from clumping of the fibers (Figure 8). Figure 9 shows a by product of the coating process, balls of coating, in this case cobalt, impregnated with reinforcement. They are prevalent on fracture surfaces since they act as defects. Another defect found on fracture surfaces was cracked cobalt particles. They are free particles from the coating process. These defects reduce the apparent toughness of the composite which is coupled with the strength and elongation to failure [8]. Careful processing and handling can remove most of these defects, increasing the strength and elongation to failure.

Dimples cover the fracture surfaces of the unrienforced matrix as shown in Figure 10 (a & b). These dimples are produced during ductile failure: void nucleation, growth and rupture. The heat treated samples have larger dimples since the voids nucleate at stress concentrators such the strengthening dispersoids which nhave coarsened during the thermal exposure.

As the test temperature rises the size of the dimples increases (Figure 10, c). The increased kinetics at elevated temperatures cause the voids to grow more rapidly. This rapid growth suppresses further nucleation of voids, so they grow larger.

Figure 11 shows that many of the dimples on the composite fracture surfaces contain fiber ends. This would usually indicate the shear strength of the interface and matrix are great enough to prevent the fiber from pulling out after fiber failure. However, these fibers are so short (< critical aspect ratio) they cannot be loaded to failure. The fiber ends act as stress concentrators for void nucleation [9]. The matrix did exhibit the dimples from the intragranular ductile failure, even at elevated temperature (Figure 12). This is in contrast to elevated temperature composite failure of precipitation hardened aluminum matrices such as 2124. A composite failure at 400oC in Figure 13 shows the intergranular failure of the composite, and the mass pull out of fibers. The 2124 alloy is so weak at this temperature it cannot transfer load to even long fibers [7].

As stated earlier the shear strength of aluminum alloys fall off rapidly above 350°C. The table in Figure 14 has the shear strength of the matrix at room and elevated temperature. Although the shear strength drops off with temperature, the matrix still has enough strength to transfer load to fibers. Included in the table are the fiber critical aspect ratios calculated using the matrix shear strength and assuming the interface is strong [10]. Obviously the critical aspect ratio goes up with temperature. The fibers in the extruded matrix are shorter than even the room temperature critical aspect ratio. If the fibers had an aspect ratio of 50 or longer, the calculated strength of the composite would be 258 MPa at 500°C [10].

Conclusions
The strength values for Al-8Fe-2Mo/Saffil composites were not particularly impressive at room temperature, but as the situation with early Silag development, gross defects controlled the strength and these can be minimized with more careful processing. The strength value obtained at 500°C (51 MPa) is encouraging and demonstrates that matrix strengthening and stability is important and controlling at temperatures above 350°C, and composite reinforcement is a valid method to increase the high temperature strength of light alloys. Development of a matrix with stable dispersions and which can be bonded, possibly aided by fiber coatings, to the reinforcements is essential to developing a high temperature aluminum composite reinforced with discontinuous fibers.

If the aspect ratio could be maintained, and the defects reduced, cobalt coated Saffil in the Al-8Fe-2Mo matrix would be a viable composite for high temperature applications.

This research was supported by AFWAL, Dr. Susan Kirchoff, Project Monitor.

References:
[1]   Starke, E.A., and Wert, J.A., "The Strengthening

Mechanisms of Aluminum Powder Alloys", <u>High Strength Powder Metallurgy ALuminum Alloys II</u>, Eds: Hildeman, G.J., and Koczak, M.J., proc. TMS-AIME: Toronto Can. Oct 13-17, 1985.

[2] Dinwoodie, W.L., and Langman, C.A.J., " 'Saffil' Alumina Fibre for Metal Matrix Composites", <u>Paper presented at UCLA Extension Conference "Metal Matrix Composites"</u>, Aug 1-5, 1983.

[3] Private communications with Dr. Henry An, Boeing Aerospace Company.

[4] Ackermann, L., Guillemin, I., Lalauze, R., and Pijolat, C., "Study of Water Desorption During Degassing", <u>High Strength Powder Metallurgical Aluminum ALloys II</u>, Eds: Hildeman, G.J., and Koczak, M.J., proc. TMS-AIME: Toronto, Can. Oct 13-17, 1985.

[5] Kim, Y-W, Griffith, W.M., and Froes, F.H., "Surface Oxides in P/M Aluminum Alloys", <u>Journal of Metals</u>, Aug. 1985, 27-33.

[6] Kirchoff, S.D., Adkins, J.Y., Griffith, W.M., and Martorelli, I.A., "Effective Method for Degassing Evaluation of Aluminum PM Alloys", <u>Rapidly Solidified Aluminum Alloys</u>, ASTM STP 890, Eds: Fine, M.E., and Starke, E.A., Philadelphia, 1986, 354-366.

[7] Schueller, R., and Wawner, F., "An Analysis of High Temperature Behavior of Discontinuously Reinforced Al Alloys", <u>Proceedings, Tenth Annual Discontinuously Reinforced MMC Working Group Meeting</u>, Park City, Utah, Jan., 1988.

[8] Griffith, W.M., and Santner, J.S., "Effect of Defects in Aluminum P/M Alloys", <u>High Strength Powder Metallurgical Aluminum Alloys</u>, Eds: Hildeman, G.J., and Koczak, M.J., proc. TMS-AIME: Dallas Tx, Feb. 17-18, 1982, 125-145.

[9] Nutt, S.R., and Needleman, A., "Void Nucleation at Fiber Ends in Al-SiC Composites", <u>Scripta Metallurgica</u>, Vol. 21, 1987, 705-710.

[10] <u>The Microstructure and Mechanical Properties of High Temperature Aluminum Composites</u>, Master's Thesis, Wm.D. Pollock, Univ. of Virginia, 1989.

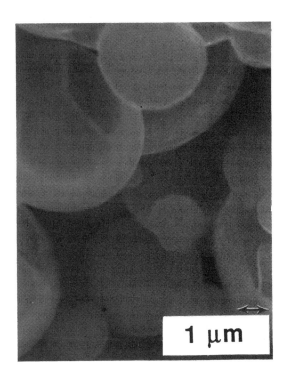

Fig 1 Rapidly Solidified
Al-8Fe-2Mo Powder.

Fig 2 Saffil,
Alumina, Fibers

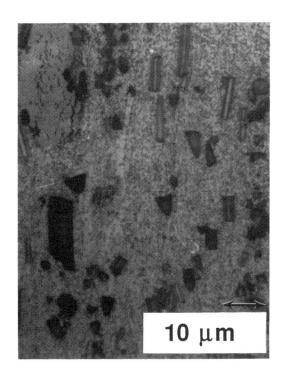

Fig 3   Extruded microstructure showing the fiber clumps, matrix
rich regions, and short fiber aspect ratio.

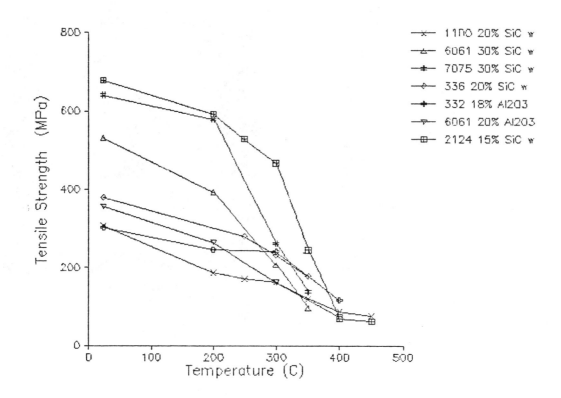

Fig 4 Strengths of various aluminum composites. The strengths converge above 350°C.

Tensile Strength of Al-8Fe-2Mo Systems

Temperature Celcius	Plain AR	Plain HT5	Saffil AR	Saffil HT5	Saffil HT4	Co/Saffil AR	Co/Saffil HT5	Co/Saffil HT4
RT	353	298	355	317	329	323	358	371
400	86	113	69	84	106	114	120	139
450	58	96	52	62	85	100	72	85
500	37	40	32	46	45	46	52	51

All strengths in MPa.

Fig 5   Table of the strength of each system at each test temperature.

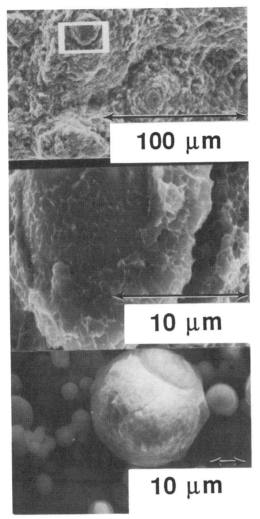

100 μm

10 μm

10 μm

Fig 6 Inclusion of Cr and Y on Fracture Surface and in the As Received Powders.

100 μm

Fig 7 Defect on Fracture Surface from Shot in Saffil.

100 μm

Fig 8 A clump of Saffil acting as a defect on a fracture surface.

100 μm

10 μm

Fig 9 Balls of Cobalt with fibers from coating process.

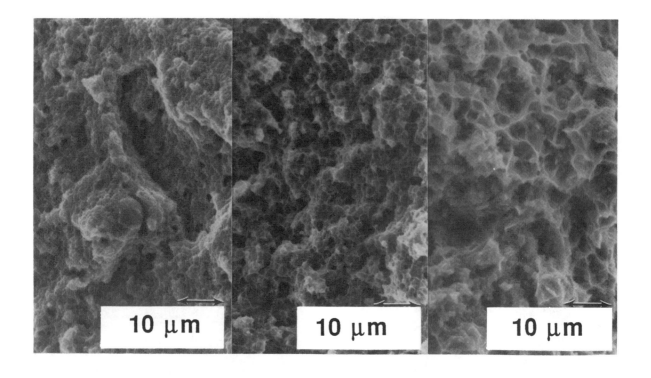

a)   b)   c)

Fig 10 Fracture surfaces of unreinforced matrix at a) room temperature as received (note the Cr and Y defect), b) room temperature HT5, and c) 500°C as received. This shows the change in dimple size with heat treatment and test temperature.

Fig 11   Co/Saffil composite as received and tested at room temperature. Note the fiber ends and dimpled matrix. The right micrograph also has a brittle cobalt defect, a by product of the coating process.

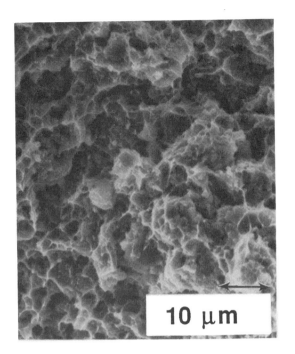

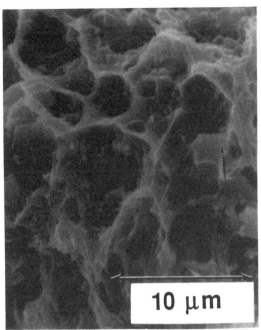

Fig 12 Cobalt coated Saffil composite tested at 500 °C. Note the lack of fiber pull out, and the large dimples even at elevated temperature.

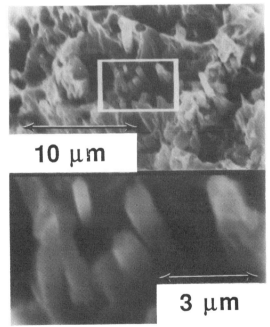

Fig 13 2124 Aluminum composite with SiC$_w$ tested at 400°C. Note the fiber pull out and failure of the matrix.

Shear Strength and Calculated Critical Aspect Ratio				
Temperature	As Received		HT5	
Celcius	Shear	Crit. Asp.	Shear	Crit. Asp.
23	185	5.4	166	6
400	53	19	60	24
500	22	46	23	44

All strengths in MPa.

Fig 14 Shear strength and calculated critical aspect ratio for various temperatures.

71

Interfacial Micromechanics of Hybrid Metal Matrix Composites

T. M. TAN, A. C. W. LAU AND A. RAHMAN

ABSTRACT

Lamina-hybrid metal matrix composites (hybrid MMC) in which adjoining matrices have different plastic responses have potential to offer more attractive damage tolerance properties than mono-matrix MMC. At the interface of two dissimilar metal matrices, both capable of deforming plastically with strain hardening, the accentuated free edge stresses may cause inter-matrix decohesion to form an interface crack. In this paper, a local-global analysis to determine the mechanics environment governing onset of such inter-matrix debonding is presented. First, an effective method of asymptotic local analysis for the singular interfacial stress in hybrid MMC is summarized and new results are presented. Then, the generalized stress intensity factor which scales the free edge interfacial stresses is solved by a global analysis for several illustrative cases. The global analysis is performed via elastic-plastic finite element method. The solutions presented completely characterize the dominant mechanical environment that governs the onset of inter-matrix decohesion. The solutions, with proper interpretation, are also applicable to hybrid MMC at elevated temperatures with matrices deforming by power-law creep.

1. INTRODUCTION

Metal matrix composites can be designed with hybrid (i.e. dual) metallic matrices (Fig. 1) to enhance damage tolerance [1]. At the free edge interface of two dissimilar metal matrices, the locally singular stresses could initiate interfacial debonding. The resulting interfacial crack may grow and ultimately lead to ply delamination. This paper addresses the the first step of this interfacial damage process. The micromechanics governing onset of free edge interfacial decohesion in hybrid-matrix MMC is solved using a local-global analysis procedure. First, the dominant singular stress field at the inter-matrix interface of hybrid MMC is determined by a local analysis employing deformation theory plasticity, asymptotic expansion and numerical shooting with parameter tracking. Then a global analysis with elastic-plastic finite element method produces the full field solution. The generalized stress intensity factor K, which links the remote field to the local field, can be determined by proper matching of the local analysis results with global analysis results.

In Section 2, the recently developed local asymptotic analysis for free edge interfacial stresses in hybrid MMC [2,3] is summarized and some newly obtained results which cover

T. M. Tan, A. C.W. Lau and A. Rahman are assistant professor, associate professor and graduate student respectively, Department of Mechanical Engineering and Mechanics, Drexel University, Philadelphia, PA 19104 essentially the entire spectrum of matrix combinations (matrix hybridizations) are presented. In Section 3, the global analysis with finite element method is described and results are presented for several illustrative cases. The solutions are examined in light of their effect on onset of decohesion of the inter-matrix interface.

2. LOCAL ANALYSIS

2.1 Formulation

Since the onset of matrix decohesion at the free edge interface of hybrid MMC is an extremely localized event (at point A, see Fig. 2), the effect of fibers, which are at distances far away comparing to the size of initial decohesion region, may be neglected. When there is stress singularity at the free edge interface, the local deformation will always be plastic, regardless of whether remote deformation in the composite is elastic or plastic. A formulation developed recently [2] for a general geometry of two bonded, plastically-deforming wedges with arbitrary angles, δ and $-\alpha$, is used for the local analysis (Fig. 2). By specifying particular values for δ and α, one may study the interfacial stress distribution at a crack tip, a notch, or at the free edge interface. In this paper on free edge interface, both δ and α have been set to be $\pi/2$.

In the plastic range, the strain hardening behavior of each metal matrix can be modelled by a power-law,

$$\tau = A_1 \gamma^{N_1} \qquad \text{for wedge 1} \qquad (1)$$

and

$$\tau = A_2 \gamma^{N_2} \qquad \text{for wedge 2} \qquad (2)$$

where τ is shear stress, γ is plastic shear strain, and A_1, A_2, N_1, and N_2 are material constants. For many metals, both A and N can be altered by thermo-mechanical treatment. The strain hardening exponent N can vary from 1 to 0, corresponding to linearly strain hardening and non-strain-hardening (perfectly plastic) respectively. Most engineering metals behave as power-law strain hardening, thus have N values somewhere between 1 and 0.

In designing hybrid MMC, one may tailor the two matrix materials to have significantly different yield strengths, but about the same strain hardening exponent. For matrices with the same hardening exponent N but different yield strengths, Eqns. (1) and (2) become

$$\frac{\tau}{\tau_{02}} = \beta \left(\frac{\gamma}{\gamma_0} \right)^N \qquad \text{for wedge 1} \qquad (3a)$$

and

$$\frac{\tau}{\tau_{02}} = \left(\frac{\gamma}{\gamma_0} \right)^N \qquad \text{for wedge 2} \qquad (3b)$$

where τ_0 and γ_0 are the reference stress and strain, and parameter β is the ratio of initial yield strengths of the two matrices, given by

$$\beta = \frac{A_1}{A_2} = \frac{\tau_{01}}{\tau_{02}} \tag{4}$$

A plane strain boundary value problem can be formulated using deformation theory plasticity in which Eqns. (3a) and (3b) are generalized to multi-axial form in terms of the equivalent shear stress τ_e and equivalent shear strain γ_e, defined as

$$\tau_e = \sqrt{\frac{s_{ij}\,s_{ij}}{2}} \quad \text{and} \quad \gamma_e = \sqrt{2\,e_{ij}\,e_{ij}} \tag{5}$$

where s_{ij} and e_{ij} are deviatoric stress and strain components, respectively. By replacing the stress and strain components in Eqns.(3a) and (3b) with the equivalent shear stress and strain, and using the relation

$$\frac{s_{ij}}{e_{ij}} = 2\left(\frac{\sigma_e}{\gamma_e}\right) \tag{6}$$

the generalized constitutive behavior can be obtained [2].

To solve for the dominant interfacial stress singularity, the stress components are expressed in a power series consisting of separable r-dependent and θ-dependent parts,

$$\sigma_{ij}(r,\theta) = r^\lambda \hat{\sigma}_{ij}(\theta) + \ldots \tag{7}$$

For the strongest stress singularity, one seeks the most negative λ within an admissible range where the strain energy and displacements are bounded as $r \to 0$. The governing equations are formulated via two stress functions $\Phi_1(r,\theta)$ and $\Phi_2(r,\theta)$ and two stream functions $\Psi_1(r,\theta)$ and $\Psi_2(r,\theta)$ for the two matrix metals [2,4]. From Eqn. (7) it is easy to see that these functions must be of the following forms

$$\Phi_k(r,\theta) = r^{\lambda+2} f_k(\theta) \qquad k = 1,2 \tag{8}$$

and

$$\Psi_k(r,\theta) = r^{\frac{\lambda}{N}+2} g_k(\theta) \qquad k = 1,2 \tag{9}$$

where k=1, or 2 denotes matrix 1 or 2, respectively. It can be shown that the condition of mechanical equilibrium for this problem is described by a pair of coupled second order differential equations in terms of $f_1(\theta)$ and $f_2(\theta)$ and their θ derivatives. This is an eigenvalue, two-point boundary value problem with λ being the eigenvalue and $\{f_k(\theta), f_k'(\theta), f_k''(\theta), f_k'''(\theta)\}$ being the associated eigenvector.

For a specific matrix combination (β and N), this eigenvalue two-point boundary value problem is numerically solved by a shooting method with parameter tracking. From solution of the eigenvalue and associated eigenvector, the radial dependence of the stresses, (r^λ), can determined, and the angular dependence of the stresses are given by proper combinations of $f_k(\theta)$, $f_k'(\theta)$, $f_k''(\theta)$, $f_k'''(\theta)$. More details of the theoretical formulation can be found in [2,3].

It is noted that the governing equations are homogeneous in $f_k(\theta)$ and its θ-derivatives. If $\{f_k(\theta), f_k'(\theta), f_k''(\theta), f_k'''(\theta)\}$ is a solution, so is $\{Kf_k(\theta), Kf_k'(\theta), Kf_k''(\theta,), Kf_k'''(\theta)\}$ for any multiplicative parameter K [2,4,5]. Therefore, the general solution for the singular stresses is:

$$\sigma_{ij}(r,\theta) = K\,r^\lambda\,\hat\sigma_{ij}(\theta) \qquad (10)$$

in which the multiplicative parameter K scales the amplitude of the local stress. By using the Mises equivalent stress, σ_e, which is

$$\sigma_e = \sqrt{3}\;\tau_e \qquad (11)$$

one can uniquely define the magnitude of K by normalizing the maximum value the angular variation of σ_e in

$$\sigma_e(r,\theta) = K\,r^\lambda\,\hat\sigma_e(\theta) \qquad (12)$$

to be unity. The parameter K, called generalized stress intensity factor, depends on remote applied loading, geometry and boundary conditions of a particular problem. It can be determined once the full field stress solution is known.

2.2 Stress Singularity at Free Edge Interface

The free edge stress singularity are solved for the entire range of strain hardening exponent N, $1 \geq N > 0$. This covers matrices that are linearly strain hardening, power-law strain hardening, or approaches non-hardening (perfectly plastic, N = 0). For each N value, solutions have been determined for essentially the entire practical range of yield strength ratio, β. The results are summarized in Fig. 3 which shows the variation of λ with N and β. These results clearly demonstrate that both the strain hardening exponent, N, and the ratio of the initial yield strengths, β, have significant effects on the singular nature of the interfacial stresses. For a specific value of N, the larger the β value, the more negative λ is. In other words, a larger difference in resistance to onset of plastic deformation between the two bonded materials leads to a stronger stress singularity at the free edge interface. For a specific β value, the free edge interfacial stresses are most singular for linearly strain hardening matrices (N = 1.0) and becomes progressively less singular as the matrices behave less and less strain-hardening in nature. As N approaches zero, the value of λ also approaches zero, as one would expect for perfectly plastic materials where stresses are finite and bounded by the yield stress.

Figure 4 shows the angular variation of the stress components, the Mises equivalent stress and hydrostatic pressure, for a typical matrix combination of $\beta = 3$ and N = 0.5. The angular variation of the stresses has been normalized such that the angular component of the Mises equivalent stress, $\hat\sigma_e(\theta)$, has a maximum of unity. It can be seen from Fig. 4 (and similar plots for other matrix combinations) that $\sigma_{\theta\theta}$ and $\sigma_{r\theta}$ are continuous across the interface ($\theta = 0$), as required by the continuity condition at the interface. On the other hand, σ_{rr} has different values on different sides of the interface ($\theta = 0^-$, and $\theta = 0^+$). This causes the Mises equivalent stress σ_e to be discontinuous across the interface. Worthy of note is that $\sigma_{\theta\theta}$ is positive at the interface ($\theta = 0$), which tends to open up the interface. If the strength of the interface is exceeded and the energy requirement for debonding is met, the interface will decohere and an interface crack would be formed.

3. GLOBAL ANALYSIS

3.1 Finite Element Model

The local analysis described in the previous section allows one to determine the spatial structure of the singular stress field at the free edge interface, but not the generalized stress intensity factor K (Eqn. (10)). K is a function of matrix combination (N and β), the remote applied loading and geometry. It must be determined from a full field solution which links the remote applied load to the local singular field. In this study, a global analysis is performed by finite element method to provide the full field solution.

The singular stresses (Eqn. (10)) have been determined by solving an eigen-problem in the local analysis: The eigenvalue λ defines the radial dependence of the stresses, and the associated eigenvector gives rise to the their angular variation. If the finite element full field solution can independently produce angular variation of stresses identical to that of the singularity solution, one can conclude that the finite element results has captured the same eigenvector of the singularity solution. Hence, the finite element computed stresses are of the form of Eqn. (10). One can then match the the finite element results to the singularity solution to determine K.

Since the singular stresses are asymptotically given by Eqn.(10) as $r \to 0$, K can be found by

$$K = \underset{r \to 0}{\text{Limit}} \left(\frac{\sigma_{ij}}{r^\lambda \hat{\sigma}_{ij}(\theta)} \right) \tag{13}$$

where σ_{ij} are the full field stresses at (r, θ) computed by finite element analysis, while λ and angular variation $\hat{\sigma}_{ij}(\theta)$ are from singularity solution of Section 2.

In plane strain elastic-plastic finite element analysis, the computed stresses may become fluctuating within an element due to the imposed incompressibility constraint as the material undergoes large plastic deformation. The inaccuracy in total stresses can be effectively avoided by using the deviatoric stress components or the Mises equivalent stress, which do not involve volumetric deformation [6]. In this paper, K has been determined from the Mises equivalent stress, i.e.

$$K = \underset{r \to 0}{\text{Limit}} \left(\frac{\sigma_e}{r^\lambda \hat{\sigma}_e(\theta)} \right) \tag{14}$$

From dimensional considerations, it can be shown that K is of the functional form

$$K = Q* \frac{\sigma_c}{d_c^\lambda} \qquad (15)$$

where σ_c is a characteristic stress, d_c is a characteristic length, and Q* is a dimensionless quantity hereinafter called the normalized generalized stress intensity factor. If the far-field applied traction σ_∞ and the composite ply half-thickness L (Fig. 5a) are used as the characteristic stress and length, then the normalized generalized stress intensity factor for the free edge interface is defined as

$$Q* = \frac{K}{\sigma_\infty/L^\lambda} \qquad (16)$$

In general, Q* is a function of material parameters N and β, and geometry.

Plane strain elastic-plastic finite element analyses were performed with the ABAQUS program using hybrid eight-node elements and reduced integration [7]. These elements have quadratic interpolation for nodal displacements but linear interpolation for pressure. This Lagrangian multiplier approach is equivalent to the penalty formulation [8]. In plane strain analysis, such arrangement is helpful to circumvent element locking (and associated inaccuracy in pressure) induced by incompressibility in the plastic range.

Figure 5a shows schematically the geometry and boundary conditions of the finite element analysis. The outer dimensions of the total mesh is 2L by w, with w = L, where L is the half-thickness of a matrix ply. The mesh consists of 408 elements and 1293 nodes. It is arranged in such a way that a total of 33 rings of elements fan out from the point of singularity, with angular spacing of $\Delta\theta=15°$. Figure 5b shows the details of the element arrangement. Starting from the second ring, the radial spacing of the elements are such that every 4 rings form a decade in units of r/R. Within each decade of r/R, the radial dimensions of these 4 rings are also spaced in logarithm scales. With this arrangement, it can be seen from Fig. 5b that the centers of elements in the innermost second to fifth rings have r/L of order of 10^{-8}. The constitutive response of the matrices are modeled as elastic-plastic with strain hardening in the plastic range. In the elastic range, the response of each matrix is linearly isotropic with Poisson's ratio of 0.3 and Young's modulus 10^7 psi. In the plastic range, the initial yield stresses of adjoining matrices are related by the ratio β defined in Eqn. (4). Beyond initial yielding, both matrices strain harden by power-law according to Eqns. (3a) and (3b). In these quasi-static analyses, the applied uniform remote traction σ_∞ increases linearly with time.

3.2 Finite Element Results

Before one can match the global and local solutions to determine Q*, one must first establish that the finite element results can capture the angular variation of the singular stress solution. Figure 6 shows comparisons of finite element results with the angular variation of the stresses from local analysis. This plot is for the matrix combination of $\beta= 5$ and N = 0.5. The finite element results are taken along one ring of elements at r/L < 10^{-7}. To make the comparison meaningful, the finite element solutions are scaled by equating the Mises equivalent stress σ_e, computed at $\theta = 7.5°$, to the value of $\hat{\sigma}_e(\theta)$ from the singular solution at the same angle. It can be seen from Fig. 6 that by matching just this one point for the Mises stress, the scaled finite element solutions for all stress components at all θ sampling points follow the angular variation of the singularity solution. In general, this observation remains true for other combinations of β and N so long as the material at the sampling finite element ring are stressed well beyond the elastic limit. It should be pointed out that the finite element stresses shown in Fig. 6 are the average of the stresses at four sampling Gaussian integration points within each element, and are attributed to the center of the element. This procedure was found to be effective in eliminating the minor fluctuation in the hydrostatic pressure computed at Gaussian points. Unless otherwise specified, all finite element results presented in the following are element-averaged values.

Since the constitutive relation used in finite element analysis is elastic-plastic, the results will always reflect certain compressibility in deformation. The singularity solution, however, is derived based on the fully plastic (hence totally incompressible) assumption. In determining Q* based on Eqn. (16) one must be careful to use only those finite element results from regions where the computed stresses clearly show characteristics of the eigenvector of the singularity solution. For this purpose, two monitoring quantities, namely, the effective Poisson's ratio, $\nu_{effective}$, and an error quantity, E_θ, are introduced. For plane strain analysis, the effective Poisson's ratio, defined as

$$\nu_{effective} = \frac{\sigma_{zz}}{\sigma_{xx} + \sigma_{yy}} \tag{17}$$

gives an indication of the degree of plastic deformation. Its magnitude approaches 0.5 as the plastic strain in the material dominates over the elastic strain. The error quantity E_θ is the root of sum of squares of the difference of the angular component of Mises equivalent stress between the normalized finite element (FEM) results and singularity (SING) results, or,

$$E_\theta = \sqrt{\sum_\theta \left[\frac{\hat{\sigma}_e^{FEM} - \hat{\sigma}_e^{SING}}{\left(\hat{\sigma}_e^{SING}\right)_{MAX}} \right]^2} \tag{18}$$

For $\beta = 4$ and N = 0.4 , the error E_θ varies with $\nu_{effective}$ as shown in Fig. 7. This plot shows data for the the entire loading history. It is apparent that as the remote loading increases, the material becomes more plastically deformed, causing $\nu_{effective}$ to increase. As $\nu_{effective}$ of a material point increases, the value of E_θ decreases monotonically. This trend continues up to about $\nu_{effective} = 0.49$, beyond which the value of E_θ starts to behave quite randomly. This indicates that the finite element results at $\nu_{effective}$ around 0.49 give the most accurate portrait of the singularity solution. Fig. 8 shows a similar E_θ plot for $\beta = 6$ and N = 0.5   Results for other combinations of β and N show that the $\nu_{effective}$ giving minimum error varies from 0.49 to 0.495. To standardize for the sake of comparison, for all cases we choose finite element results at $\nu_{effective} = 0.49$ to match with the singularity solution .

From finite element results, one can evaluate the quantity Q, defined as

$$Q = \left(\frac{\sigma_e^{FEM}}{\sigma_\infty \left(\frac{r}{L} \right)^\lambda \hat{\sigma}_e(\theta)} \right) \tag{19}$$

To determine the value of Q* from Q, it is necessary to extrapolate the Q results to the limit of $r \rightarrow 0$ according to continuum mechanics theory. One realizes that for a typical ply thickness 2L=0.01 inch, a point at a distance six orders of magnitude smaller than the ply half-thickness (i.e. $r/L = 10^{-6}$) away from the free edge interface is about one atom diameter away from the origin. For good measure, we chose $r/L \doteq 10^{-8}$ (corresponding to an r of subatomic dimension away from the origin) as the limiting zero dimension for continuum mechanics to be relevant. Hence, finite element results at $r/L \doteq 10^{-8}$ can be considered as data at the limit of $r \rightarrow 0$. In this paper Q* is determined form the value of Q at $r/L \doteq 10^{-8}$ when $\nu_{effective} = 0.49$, or,

$$Q^* = \left(\frac{\sigma_e^{FEM}}{\sigma_\infty \left(\frac{r}{L} \right)^\lambda \hat{\sigma}_e(\theta)} \right)_{\substack{\nu_{effective} = 0.49 \\ \frac{r}{L} \doteq 10^{-8}}} \tag{20}$$

Results of the normalized general stress intensity factor, Q*, covering a practical range of N and β combinations, are summarized in Fig. 9. Since angular variation of the finite element stresses agrees very well with that of the singularity solution, Q* determined using elements at different θ gives essentially the same value. The results presented in Fig. 9 are from the element closest to the interface.

5. SUMMARY AND CONCLUSIONS

The free edge interfacial stresses that govern onset of interface decohesion in hybrid metal matrix composites have been solved by a local-global analysis.

(1) For hybrid matrix composites deforming plastically with the same strain hardening exponent N but different yield strengths, the local analysis indicates that the stresses at the free edge inter-matrix interface have a power-type singularity of the form $\sigma_{ij} = Kr^\lambda \hat{\sigma}_{ij}(\theta)$. The strength of this stress singularity, λ, is affected by both the strain hardening exponent N and the ratio of the initial yield strengths β. The variation of λ for the entire range of β and N combinations, as summarized in Fig. 3, agrees with physical insight and obey known requirements at limiting cases.

(2) The normalized generalized stress intensity factor Q*, defined in Eqn. (16), can be effectively obtained from results of a global elastic-plastic finite element analysis. This analysis uses the eigenvector as a vehicle to screen for the optimal global solution to match with singularity solution from local analysis, and uses a procedure to circumvent fluctuations in hydrostatic pressure.

(3) The spatial structure of the singular stresses determined by the local analysis, and Q* determined by the local-global matching, together completely characterize the dominant

mechanics environment that controls onset of free edge inter-matrix debonding in hybrid MMC. After the onset of decohesion, the resulting interfacial crack may be analyzed in the framework of interfacial crack mechanics in elasto-plastic media [9].

(4) When displacements and strains in the analysis are interpreted as their time rates, these solutions are applicable at elevated temperatures where the matrices deform by power-law creep [6].

Acknowledgements --- The work of ACWL was partially supported by the Strategic Defense Initiative Office/Innovative Science and Technology through the Solid Mechanics Program, Office of Naval Research, contract No. N00014-85-K-0628 with Dr. Y. Rajapakse as program monitor. Some computations were performed with NSF supercomputer time. The ABAQUS finite element code was made available under academic license from Hibbit, Karlsson, and Sorensen, Inc., Providence, Rhode Island.

REFERENCES

1. Awerbuch, J., and Koczak, M. J., "Experimental and Micromechanics Studies of Metal Matrix Composites," Drexel University, private communications, 1987.
2. Lau, C. W., and Delale, F., " Interfacial Stress Singularities at Free Edge of Hybrid Metal Matrix Composites," ASME Journal of Engineering Materials and Technology, Vol. 110, 1988, pp. 41- 47.
3. Lau, C. W., Delale , F., and Rahman, A., " Interfacial Mechanics of Seals," in Technology of Glass, Ceramic, or Glass-Ceramic to Metal Sealing, MD-Vol. 4, pp. 89-98, ASME, New York, 1987
4. Rice, J.R. and Rosengren, G.F., "Plastic Strain Deformation Near a Crack Tip in a Power-Law Hardening Material," Journal of the Mechanics and Physics of Solids, Vol. 16, 1968, pp. 1-12.
5. Hutchinson, J.W., "Singular Behaviour at the End of a Tensile Crack in a Hardening Material," Journal of the Mechanics and Physics of Solids, Vol. 16, 1968, pp. 13-31.
6. Lau, C. W., Argon, A.S., and McClintock, F.A.," Application of Finite Element Methods in Micromechanical Analyses of Creep Fracture Problems," Computers and Structures, Vol. 17, No. 5-6, 1983, pp. 923 - 931.
7. ABAQUS Theory Manual, Hibbit, Karlsson & Sorensen, Inc., Providence, Rhode Island, 1987.
8. Nagtegaal, J. C., Park, D. M. and Rice, J. R., "On Numerically Accurate Finite Element Solutions in the Fully Plastic Range," Computer Methods in Applied Mechanics and Engineering, Vol. 4, 1974, pp. 153-178.
9. Shih, C. F., and Asaro, R. J., "Elastic-Plastic Analysis of Cracks on Bimaterial Interfaces: Part I: Small Scale Yield," ASME Journal of Applied Mechanics, Vol. 55, 1988, pp. 299-316.

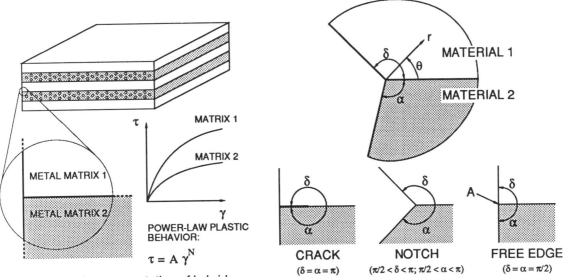

Fig.1 Schematic representation of hybrid
metal matrix composites

POWER-LAW PLASTIC
BEHAVIOR:

$\tau = A \gamma^N$

Fig.2 Geometries of bonded bimaterial wedges

CRACK
($\delta = \alpha = \pi$)

NOTCH
($\pi/2 < \delta < \pi$; $\pi/2 < \alpha < \pi$)

FREE EDGE
($\delta = \alpha = \pi/2$)

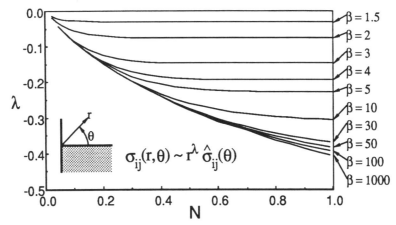

$\sigma_{ij}(r,\theta) \sim r^\lambda \, \hat{\sigma}_{ij}(\theta)$

$\beta = 1.5$
$\beta = 2$
$\beta = 3$
$\beta = 4$
$\beta = 5$
$\beta = 10$
$\beta = 30$
$\beta = 50$
$\beta = 100$
$\beta = 1000$

Fig.3 Solution of λ for free-edge interfaces

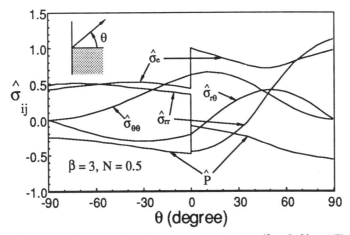

$\hat{\sigma}_e$

$\hat{\sigma}_{r\theta}$

$\hat{\sigma}_{\theta\theta}$

$\hat{\sigma}_{rr}$

$\hat{P}$

$\beta = 3$, $N = 0.5$

Fig.4 Angular variation of stress components ($\beta = 3$, $N = 0.5$)

81

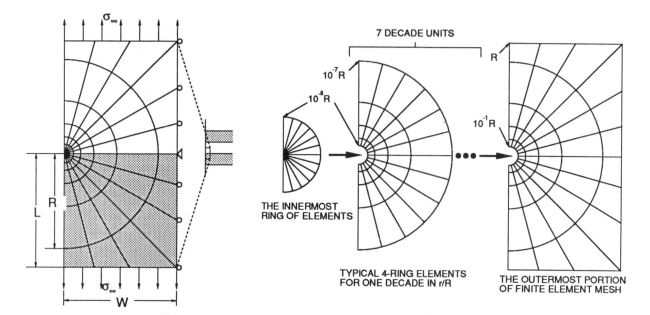

Fig.5a Schematic representation of
finite element model

Fig.5b Details of finite element mesh

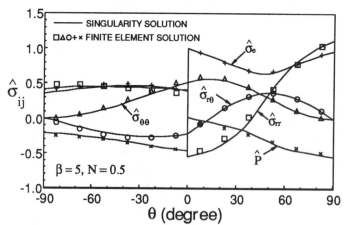

Fig.6 Comparison of stress angular variations ($\beta = 5$, $N = 0.5$)

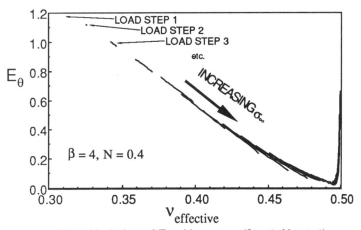

Fig.7 Variation of E_θ with $\nu_{effective}$ ($\beta = 4$, N = 0.4)

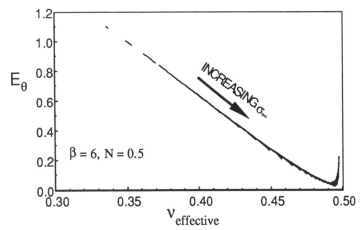

Fig.8 Variation of E_θ with $\nu_{effective}$ ($\beta = 6$, N = 0.5)

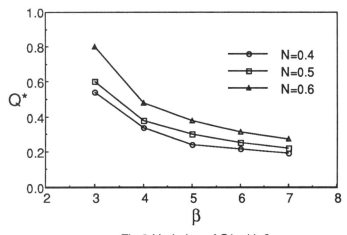

Fig.9 Variation of Q* with β

Processing and Failure Mode of Al/SiC$_w$ MMC

T. LIM, S. Y. LEE AND K. S. HAN

ABSTRACT

This study deals with fabrication and microstructural analysis of MMC. SiC reinforced 6061 Al alloy composites have been fabricated by the modified compocasting. Processing variables such as mixing temperature, stirring time, types of whisker introduction, whisker weight fraction, etc., are examined. Particular attention is focused on optimizing time of processing variables under the open atmosphere.

Energy dissipating capacity of MMC slightly depends on heat treatment and fabrication conditions. After solution treatment, charpy impact test was carried out at room and elevated temperatures. It is shown that total impact energy dissipated of MMC is much less than that of unreinforced Al alloy. Also, effect of weight fraction of whiskers on impact energy is studied.

Metallurgraphic examination shows that whiskers are distributed homogeneously but there are some agglomerations due to the shape of reinforcements. Fracture surface is examined using SEM. It is shown that failure mode is a brittle one due to poor bonding between whiskers and matrix, the presence of oxide and intermetallic compounds, voids, high dislocation density region adjacent to whiskers, and non-wetted zones. Effects of temperature and weight fraction of whisker on failure mode are also studied. And the failure mechnism of short fiber (whisker) composites is also proposed.

INTRODUCTION

Discontinuous fiber reinforced metal matrix composites (MMCs) are being developed due to not only high specific mechanical properties at room and elevated temperatures but the ability to be processed by the conventional methods. Furthermore, these MMCs exhibit better ductilities than continuous ones because of some plastic flows near fiber ends.

Powder metallurgy, squeeze casting, rheocasting (compocasting), and spray casting, etc., are available for the fabrication of discontinuous MMCs [1-3]. Among them, compocasting is suitable for the low-cost mass-production and leads its products to near net shape. But, compocasting has some disadvantages; restriction on weight fractions of reinforcements, a brittle failure because of segregations of reinforcements (when reinforcements are whiskers or fibers, they are likely to be located in interglobular regions,

T.Lim ; Graduate student of Dept. of Mech. & Aerospace Eng., State University of New York at Buffalo, Amherst, NY, 14260.
S.Y. Lee and K.S. Han ; Graduate student and Associate professor of Dept. of Mechanical Eng., Pohang Institute of Science and Technology, Pohang, Korea.

which cause a brittle failure), difficulty of introducing reinforcements to the melt without clustering, breakage of reinforcements due to the shearing effect of slurry and the rotation of impeller [4-7].

In this study, discontinuous MMCs reinforced by SiC whiskers have been processed using modified compocasting technique under various processing conditions which include mixing temperature, introduction types of reinforcements, stirring time, and fiber weight fractions. The mechanical behaviours have been examined through static and dynamic tests to determine the stiffness and energy dissipating capacity. Effects of processing variables and testing temperatures on the mechanical behaviours have been studied to optimize the processing conditons for modified compocasting. Failure modes at different temperatures (from room temperature to 300°C) and failure mechanisms of Al / SiC$_w$ MMCs have also been examined by fractographies from SEM.

EXPERIMENTAL PROCEDURE

The Al alloy used in this study was 6061 Al alloy. Reinforcements were SiC whiskers (American Matrix Inc.) and properties of whiskers are shown in Table 1. Discontinuous MMCs were fabricated by the modified compocasting under the open atmosphere and the apparatus for modified compocasting is shown schematically in Fig. 1.

The alloy was melted in the electric furnace. The temperature of melt at the upper region was set about 630 ~ 640°C and that of the lower one was just above the liquidus temperature, 660°C. These temperatures were kept throughout the compositing cycle and could be achieved by controlling the convection of the upper melt and the power of the furnace. The stainless steel stirrer coated by the china clay was located just below the surface of the melt and driven by variable speed D.C. motor. For the addition of whiskers to the melt, a method using the feeder under the open atmosphere was used. After stirring for the presetting time, the molten metal was remelted to recover its fluidity, and then poured into permanent molds.

To examine effects of heat treatment on mechanical behaviours, samples were heat treated, T$_6$(solution treatment at 540°C for 4 hours, water quenching, aging at 180°C for another 4 hours). Tensile tests for stiffness and charpy impact test were carried out [8,9]. For high temperature tests, specimens were kept at desired temperatures for 1 hour. Fracture surfaces and microstructures were observed through SEM to study the failure mode, distribution of whikers, and bonding between reinforcements and matrix. Microstructures were examined after etching with 30% hydrochloric acid.

RESULTS & DISCUSSION

1, Fabrication of composites

SiC whiskers were successfully dispersed in 6061 Al alloy by the fabrication method described above. Using the modified compocasting with the feeder under the open system, it is possible to introduce up to about 9 wt% of whiskers. When the amount of whiskers added to the melt were more than 9 wt%, whiskers were severely rejected by the melt. Typical microstructure of Al / SiC$_w$ composites is shown in Fig 2. This shows relatively uniform distribution of whiskers, and interfacial compounds formed near whiskers.

On the other hand, agglomerations of whiskers were observed in materials using the feeder under the open atmosphere. The amount of agglomerations could be reduced by the agitation. As longer the stirring time is, the number and size of agglomerations are getting reduced to some extent. However, there is a limitation for reducing agglomeratons by the agitation only. It is prop osed that the amount of agglomeration can be reduced remarkably by the injection method for charging reinforcements with high pressure of nitrogen gas.

To optimize the stirring time, the distribution of whiskers and the amount of interfacial compounds have to be considered. As the stirring time increases, the distribution is getting better but excessive amount of interfacial compounds lead to a brittle failure. From the impact data and morphology, it is proposed that the optimum stirring time is 5 minutes after the complete introduction of whiskers.

In the ordinary compocasting, whiskers can not be dispersed in the partially solidified slurry. It causes segregation of whiskers and brittle failure. In the modified compocasting, because the amount of slurry is less than that in the compocasting and the possibility of whiskers to be broken during the compositing is reduced, sound composite materials can be fabricated.

2, Mechanical properties of composites

Impact energy of discontinuous MMCs is usually lower than unreinforced matrix. The small strain to failure and high stiffness of whiskers confined the plastic flow of ductile matrix so that the fracture of composites exhibit brittle characteristics [10,11]. To measure impact energy, various kinds of specimens were prepared. Fig. 3 shows the variation of impact energy according to the weight fraction of whiskers. Both in heat treated and as fabricated specimens, impact energy is lowerd as the weight fraction increases. The effect of whisker content on impact energy of composites depends on the energy dissipating capacity of whiskers and matrix. There are slight differences of impact energy between heat treated and as fabricated specimens. Heat treated ones show little lower impact energy than as fabricated ones.

Fig. 4 shows the relations between impact energy and stirring time. As the weight fraction of whiskers changes, the optimum stirring time to get maximum impact energy changes. Therefore, as mentioned above, the optimum stirring time is the additional 5 minutes after the complete introduction of reinforcements.Some specimens show high impact energy (about two times higher energy than the unreinforced matrix). By the examination of their fracture surfaces, multiple cracking, fiber pull-out, fiber debonding, and bridging of fibers are found. These cause a high impact energy because of their contribution to the increase of crack propagation energy.

Notch sensitivity of MMCs is another important factor in its properties. The comparison of impact energy between notched and unnotched specimens is shown in Table 2. Impact energy of notched ones is slightly lower than unnotched ones. But the energy per unit area of both are the same. So, it is considered Al / SiC$_w$ MMCs are not notch sensitive.Impact energy at elevated temperatures increases due to the transition of matrix to the ductile mode. The effect of testing temperature on the impact energy is well explained in Fig. 5.

By the addition of reinforcements, Al / SiC$_w$MMCs have the improvements in stiffness over the unreinforced alloy. Comparisons between experimental data and theoretical values [12,13] are shown in Table 3. Experimental data have lower values than theoretical ones on account of imperfect bondings, defects (cavities & precipitates) of matrix, non-wetted zones,and breakage of whiskers during the fabrication. As the weight fraction increases, differences between data and expected values also grows, because of greater possibilities of these defaults inside of composites at high weight fractions.

3, Failure analysis of composites

The fracture surfaces of the impact specimens were examined to determine the failure mode and the failure mechanism. Fig.6 shows typical fracture surfaces of Al / SiC$_w$MMCs. Matrix shows a brittle failure adjacent to whiskers, but some portions of matrix show a localized ductility.

Matrix is changed to be brittle after heat treatment because of precipitation hardening. In discontinuous MMCs, the distribution and bonding condition of reinforcements are more determing factor for changing the failure mode than brittle matrix. The effect of heat treatment on the amount of interfacial compounds is negligible under the normal condition. Under the overaged condition, the amount of interfacial compounds is increased, then MMCs are getting more brittle[14-16]. Therefore, heat treatment has negligible effect on the failure mode.

Fracture at agglomerations is illustrated in Fig.7. Cracks are initiated and propagated through agglomerations with a brittle manner. Only few whiskers show a fiber pull-out and bridging and most whiskers are non-wetted, therefore, fracture at agglomerations represents brittleness. Failure of discontinuous MMCs shows a brittle mode due to the low failure strain of reinforcements, cavities in matrix, and non-wetted reinforcements which acts as a precrack.

Cavities of matrix are due to the shrinkage during the solidification [17]. Cracks are initiated from imperfections (cavities of matrix and agglomeratios of whiskers) and high dislocation density regions (ends and interfaces of whiskers). In discontinuous MMCs prepared by casting methods, agglomerations are also very critical sites for the fracture initiation. After the initiation, matrix cracking propagates along the preferred region (where stress concentration in matrix is the greatest). Crack propagations along the whisker end and interfaces between intermetallic compounds and matrix are shown in Fig.8. With propagating matrix cracks, fracture stress is decreased, therefore, a catastrophic failure occurs. At elevated temperatures, failure mechanism is basically similar to that at room temperature, but dimples in matrix are getting deeper and areas of the localized ductility increase, indicating ductile failure mode of matrix as a temperatue increases.

CONCLUSIONS

The following comments are made:

1, Al / SiC$_w$ MMCs can be made by the modified compocasting. And a homogeneous disribution and sound composite materials with less segregations and reduced fiber breakages were obtained by a proposed fabrication method.
2, Optimum stirring time is 5 minutes after the complete introduction for the homogeneous distribution.
3, Impact energy decreases with the weight fraction of whiskers due to the low failure strain and high stiffness of whiskers which confine the plastic flow of matrix. The impact energy at elevated temperatures increases because of the transition of matrix from brittle mode to ductile mode. Fiber debondings, pull - out, bridging of fibers should be induced and agglomerations and cavities of matrix also minimized to improve impact properties.
4, Stiffness can be improved by the addition of reinforcements. Better improvement in stiffness can be achieved by reducing defects inside composites.
5, Failure mode of Al / SiC$_w$ MMCs is brittle, but some localized ductile matrix can be found. Cracks used to be initiated from cavities of matrix and ends of reinforcements and propagate along the high dislocation density region.

REFERENCES

1, A.P. Divecha, S.G. Fishman, and S.D. Karmarkar, " Silicon carbide reinforced aluminum - A formable composite ", J. of Metals, Sept., 1981, pp. 12 ~ 17.
2, T.W. Chou, A.Kelly, and A. Okura, " Fibre-reinforced metal matrix composites ", Composites, Vol. 16, No. 3, July, 1985, pp. 187 ~ 206.
3, F.A. Girot, J.M. Quenisset, and R. Naslain, " Discontinuously-reinforced aluminum matrix composites ", Composites Science and Technology, 30, 1987, pp. 155 ~ 184.

4, Deonath, R.T. Bhat, and P.K. Rohatgi, " Preparation of cast aluminium alloy-mica particle composites ", J. of Materials Science, 15, 1980, pp. 1241 ~ 1251

5, M.K. Surappa and P.K. Rohatgi, " Preparation and properties of cast aluminium - ceramic particle composites ", J. of Materials Science, 16, 1981, pp. 983 ~ 993.

6, B.F. Quigley, G.J. Abbaschian, R. Wunderlin, and R. Mehrabian, " A method for fabrication of aluminum - alumina composites ", Metallurgical Transaction A, Vol. 13A, January, 1982, pp. 93 ~ 100.

7, F.M. Hosking, F. Folgar Portillo, R. Wunderlin, and R. Mehrabian, " Composites of aluminum alloys: fabrication and wear behaviour ", J. of Materials Science, 17, 1982, pp. 477 ~ 498.

8, T. Lim and K.S. Han, " The effective stiffness of the random fiber composites ", to be published, ICCM - VII, Beijing, China, 1989.

9, R.M. Christensen and F.M. Waals, " Effective stiffness of randomly oriented fiber composites ", J. of Composites Materials, Vol. 6, 1972, p. 518.

10, K.M. Prewo, " The charpy impact energy of Boron-aluminum ", J. of Composite Materials, Vol. 6, Oct. 1972, pp. 442 ~ 456.

11, C.M. Friend and A.C. Nixon, " Impact response of short δ - alumina fibre/aluminium alloy metal matrix composites ", J. of Materials SCience, 23, 198, pp. 1967 ~ 1973.

12, " Notched bar impact testing of metallic materials ", ASTM Standards Part 10, E - 23, Philadelphia, PA, 1978.

13, " Test method for tensile properties of fiber reinforced metal matrix composites ", ASTM Standards for Composite Materials, D - 3552 - 77 , Philadelphia, PA, 1987.

14, A. Skinner, M.J. Koczak, and A. Lawley, " Work of fracture in aluminum metal matrix composites ", Metallurgical Transaction A, Vol. 13A, February, 1982, pp. 289 ~ 297.

15, T. Christman and S. Suresh, " Microstructural development in an aluminum alloy - SiC whisker composite ", Acta Metallurgica, Vol. 36, No. 7, 1988, pp. 1691 ~ 1704.

16, H.J. Rack and R. Ratnaparkhi, " Damage tolerance in discontinuously reinforced metal-matrix composites ", J. of Metals, Nov., 1988, pp. 55 ~ 57.

17, A. Kohyama, N. Igata, Y. Imai, Teranishi, and T. Ishikawa, " Microstructures and mechanical properties of silicon-carbide fiber reinforced aluminum composite materials and their preform wires ", ICCM - V, Edited by W.C. Harrigan, Jr., J. Strife, and A.K. Dhingra, Metallurgical Society of AIME, Warrendale, PA, 1985, pp. 609 ~ 621.

Table 1 ; Mechanical and physical properties of SiC whisker

Elastic modulus (GPa)	:	350 - 380
U T S (MPa)	:	5000 - 7000
Ductility (%)	:	< 1
Chemistry	:	Stiochiometric SiC
Strucuture	:	Beta Phase (Cubic)
Impurity	:	Less than 1000 ppm
Diameter	:	1 - 3 μ
Length	:	30 - 200 μ
Specific gravity	:	3.21

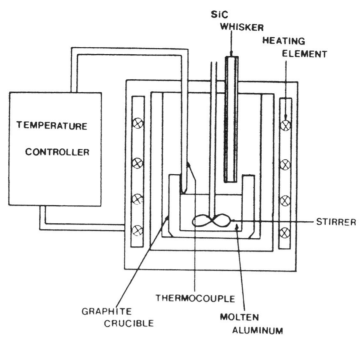

Fig. 1; The apparatus for the modified compocasting.

Table 2 ; Effect of notch on impact energy of 3 wt% SiC_w / Al specimens.

Specimen	Energy (J)	Energy per unit area (J / cm^2)
Notched	3.83	4.78
Unnotched	4.85	4.85

Table 3 ; Comparisons of stiffness between experimental data and
theoretical values

	Experimental data	Theoretical values
Al 6061 T6	69 GPa	68.9 GPa
3wt% Al/SiC	73.35 GPa	74.85 GPa
6wt% Al/SiC	78.4 GPa	80.7 GPa
9wt% Al/SiC	80.6 GPa	86.5 GPa

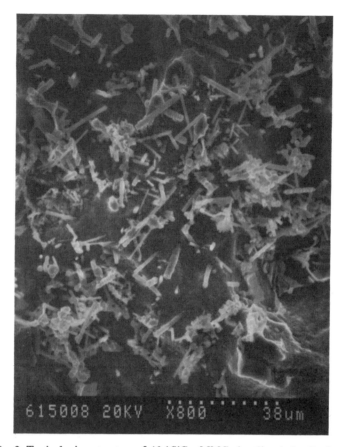

Fig. 2; Typical microstructure of Al / SiC$_w$ MMC, 6 wt%, stirring for 5 minutes

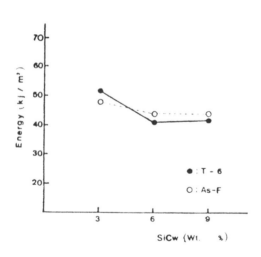

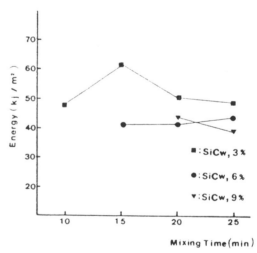

Fig. 3; Effects of weight fractions of whiskers on the impact energy.

Fig. 4; Effects of the stirring time on the impact energy.

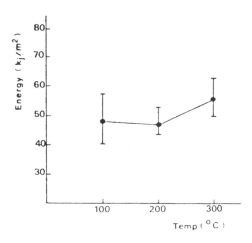

Fig. 5; Effects of testing temperatures on the impact energy.

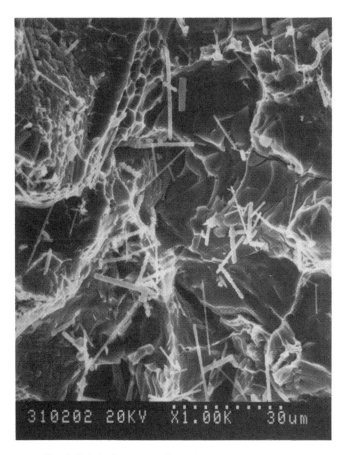

Fig. 6; Brittle fracture surface of Al / SiC$_w$ MMC,
3 wt%, stirring for 10 minutes.

Fig. 7; Fracturen at agglomerations of whiskers,
3 wt%, stirring for 15 minutes.

Fig. 8; Trends of crack propagations of Al / SiC$_w$ MMC, 6 wt%, stirring for 20 minutes.

(a) Cracks propagate along interfaces between interfacial compounds and matrix.

(b) Cracks propagate along ends of whiskers.

93

Effect of a High Temperature Cycle on the Mechanical Properties of Silicon Carbide/Titanium Metal Matrix Composites

R. A. NAIK,* W. S. JOHNSON** AND W. D. POLLOCK***

ABSTRACT

   The effects of a simulated superplastic forming/diffusion bonding (SPF/DB) temperature cycle on the tensile static and fatigue properties of a 3-ply [0/90/0] silicon carbide/titanium composite were investigated. Unreinforced Titanium was also tested to study the effect of the SPF/DB cycle on the matrix static properties. Fibers were etched from the composite and then individually tested for modulus and strength. Finally, a microscopic examination of the fiber/matrix interface was performed to study the effects of the SPF/DB cycle on the interface. The [0/90/0] composite subjected to the SPF/DB process showed a 25 percent decrease in ultimate tensile strength (UTS) and a 30 percent decrease in failure strain compared to the as-fabricated (ASF) material. The fatigue endurance limit at 50,000 cycles for the SPF/DB specimens was 50 percent lower than the ASF specimens. The fracture surface of the ASF specimens was very irregular with protruding fiber ends as compared to the planar fracture surface of the SPF/DB cycled specimens that showed fiber ends flush with the surface. The large changes in the tensile strength, fatigue life, and fracture surface appearance due to the SPF/DB cycle are explained by a difference in the failure mechanisms occurring as a result of the SPF/DB-induced changes in the strength of the fiber/matrix interface.

INTRODUCTION

   Titanium matrix composites reinforced with continuous silicon carbide (SCS-6) fibers are of current interest as structural materials for high temperature applications. Some of the potential applications of these composites are the National Aerospace Plane, aircraft engines, the Advanced Tactical Fighter and other high speed aircraft. These titanium matrix composites may be joined to a titanium substructure by a high temperature superplastic forming/diffusion bonding (SPF/DB) process. In such a process, the titanium matrix composites will be subjected to the temperature/time cycle associated with the SPF/DB process. A major concern is the effect of this thermal cycle on the matrix, the fiber and the fiber/matrix interface and the consequent effects on composite properties. The present study investigates the mechanical properties of a 3-ply [0/90/0] SCS-6/Ti-15V-3Cr-3Al-3Sn [1] (referred to as Ti-15-3) laminate after exposure to a simulated SPF/DB cycle.

   Several investigators [2,3] have previously reported relevant results for laminates subjected to a similar thermal processing cycle. Harmon, Saff, and Sun [2] studied the static properties of $(B_4C)B/Ti-15-3$ that had been subjected to a simulated superplastic forming/diffusion

* R. A. Naik, Planning Research Corporation, Hampton, Virginia 23666.
** W. S. Johnson, NASA Langley Research Center, Hampton, Virginia 23665.
*** W. D. Pollock, Resident Researcher, University of Virginia.

bonding (SPF/DB) cycle in which the composite is kept at 1000 deg C for 1 hour. They found a significant (over 35 percent) reduction in static strength as a result of the SPF/DB cycle for this boron/titanium composite. Kaneko and Woods [3] studied the effect of a simulated braze cycle (peak temperature of 690 deg C) on the static strength of unreinforced titanium. The tensile strength of the Ti-15-3 was reduced about 7 percent by the thermal processing. An understanding of the failure mechanisms and the effect of the SPF/DB cycle on the composite and constituents is currently lacking.

The objective of the present study is to investigate the effects of the SPF/DB cycle on the SCS-6/Ti-15-3 composite, the matrix, the fiber, and the fiber/matrix interface. Composite specimens were tested under static loading to study the effect of the simulated SPF/DB cycle on the modulus and strength. Unreinforced Ti-15-3 was also tested to study the effect of the SPF/DB cycle on the matrix static properties. Next, the effect of the SPF/DB cycle on the fatigue response and failure modes of the composite was investigated. Fibers removed from the two panels by dissolving the titanium matrix were then individually tested for modulus and strength. Finally, a detailed microscopic examination of the fiber/matrix interface was performed to study the effects of the SPF/DB cycle on the interface.

EXPERIMENTAL PROCEDURES

Materials and Specimens

Ti-15V-3Cr-3Al-3Sn is a metastable beta titanium alloy [1]. SCS-6 fibers are silicon carbide fibers that have a carbon core and a thin carbon-rich surface layer [4]. The typical fiber diameter is 0.142 mm. The composite laminates are made by hot-pressing Ti-15-3 foils between unidirectional tapes of SCS-6 fibers. For the present study, two panels with a 3-ply [0/90/0] layup were used. Laminate thickness was approximately 0.68 mm. The fiber volume fraction was measured to be 0.375.

One of the panels was tested in the as-fabricated (ASF) condition. A second panel was subjected to a thermal processing cycle that simulated a superplastic forming/diffusion bonding operation. This simulated SPF/DB cycle was performed in a vacuum furnace and consisted of raising the temperature from ambient to 700 deg C at a rate of 10 deg C per minute. After stabilizing at 700 deg C the temperature was further increased to 1000 deg C at a rate of 4 deg C per minute. The panel was held at 1000 deg C for 1 hour. It was then furnace cooled to 594 deg C at a rate of 8 deg C per minute and held at that temperature for 8 hours. Finally, it was furnace cooled to 150 deg C and held for about 10 hours before cooling down to ambient temperature.

Each panel was cut into 19 mm by 152 mm rectangular specimens using a diamond wheel saw. These specimens were then "dog-boned", as shown in figure 1, using electro-discharge machining. One rectangular unreinforced titanium specimen (12.7 mm by 152 mm) that had been subjected to the SPF/DB cycle was also tested. This unreinforced titanium specimen was, in fact, a "fiberless composite" and was made by consolidating Ti-15-3 foils by the same temperature-time-pressure cycle used for the composite laminates.

Specimen Testing

Tests were conducted in a servo-hydraulic test stand under load control with a loading rate of 45 N/s. The $0°$ fibers were parallel to the loading direction in all cases. The specimens were strain-gaged for moduli determination under static loading. An extensometer was used to measure the axial strain of the fatigue specimens. All fatigue specimens were tested at an R ratio of 0.1 and at a frequency of 10 Hz. A digital data acquisition system was used to collect load and stain data. All tests were performed at room temperature.

Fiber Testing

The effect of the SPF/DB cycle on the fiber properties was determined by testing single fibers from each panel. Thin strips, 3.2 mm wide, were cut from the ASF and the SPF/DB cycled panels. These strips were then dipped into a bath of hydroflouric acid to dissolve all the titanium. The bare fibers were tested individually in a screw-driven machine. A single fiber was mounted between a special C-shaped fixture made of two cardboard pieces (see figure 2) for ease of handling and gripping. After the C-fixture was gripped, the cardboard was cut so all the load would be supported by the fiber. The gage length of the fiber between the grips was measured before loading. The fiber was then tested to failure while the load and crosshead displacement were recorded on a X-Y plotter and a computer-controlled data acquisition system. The fiber diameter was measured using an optical microscope and was typically 0.142 mm. Typical gage length of the fibers tested was 75 mm. Fiber modulus and strength were determined from the recorded data.

RESULTS AND DISCUSSION

This section presents the experimental results for the composite and its constituents. First, the effects of the SPF/DB cycle on unreinforced titanium properties is presented followed by the static and fatigue properties of the composite and the observed failure modes for the ASF and the SPF/DB cycled panels. Next, a microscopic examination of the failure surfaces is presented. Then, the results of single fiber tests and the effects of the SPF/DB cycle on fiber properties is presented. Finally, a microstructural analysis of the fiber/matrix interface for the ASF and the SPF/DB cycled panels is presented.

Unreinforced Ti-15-3 properties

The Ti-15-3 properties were obtained from tests conducted on the "fiberless composite" described earlier. The stress-strain curves for the ASF, SPF/DB cycled and aged Ti-15-3 are shown in figure 3. The ASF and aged (482 deg C for 16 hours) properties were reported in reference 5. The stress-strain curve for the SPF/DB cycled specimen lies in-between the curves for the ASF and aged conditions. The SPF/DB cycled specimen, under static tension, showed a 9 percent increase in modulus and ultimate tensile strength (UTS) and a 9 percent decrease in the strain to failure as compared to the as-fabricated Ti-15-3. As compared to the aged specimens, the SPF/DB cycled specimen showed a 7 percent lower modulus. Thus the SPF/DB cycle had only a small effect on the static properties of the unreinforced Ti-15-3.

Composite properties

The mechanical properties, under static tension were determined from at least three specimens cut from each panel. The specimens cut from the SPF/DB cycled panel showed a 9 percent increase in modulus but a 25 percent decrease in UTS and a 30 percent decrease in failure strain as compared to the ASF specimens (see Table 1). Under cyclic fatigue (R = 0.1 at 10 Hz), the SPF/DB cycled specimens showed a significant reduction (over 50 percent) in the endurance limit (figure 4) as compared to the ASF specimens. The fatigue data for other layups of SCS-6/Ti-15-3 is included (from reference 6) in figure 4 for comparison. Although the ASF specimens show comparable fatigue life with other layups that have similar UTS, the SPF/DB cycled specimens showed very poor fatigue life. In order to study the reasons for the lower UTS and fatigue properties of the SPF/DB cycled specimens, a microscopic study was performed.

The fracture surfaces for the statically tested ASF and SPF/DB cycled specimens were significantly different as shown in figure 5. The ASF fracture surfaces were in general more irregular than those of the SPF/DB cycled specimens. Also the fiber pull-out from the fracture surface was much greater in the ASF specimens than in the SPF/DB cycled specimens where the fibers appeared to be almost flush with the fracture surface. The rough fracture surface on the ASF specimens indicates that the crack had to take a torturous path as it progressed through the composite debonding the fiber/matrix interface as it progressed. The relatively smooth fracture surface on the SPF/DB cycled specimens indicates that the crack did not deviate along the 0 deg fiber length but broke fibers as it went. Also, notice the split fiber in the 90 deg (transverse) ply. This split fiber was typically observed in the SPF/DB cycled specimens but not in the ASF specimens. The fracture surfaces of the fatigue specimens were similar to those observed for the static specimens from each panel type.

As discussed earlier, the matrix strength increased by about 9 percent. Interestingly the composite UTS (Table 1) dropped 25 percent and the fatigue endurance limit (figure 4) dropped by over 50 percent. The effect of the SPF/DB cycle on fiber properties and/or the fiber/matrix interface might explain these large differences.

Fiber properties

Single fiber tests were performed on fibers that were etched away from the two panels. At least three fibers were tested from each panel. UTS for fibers from the ASF and SPF/DB cycled panels were about 3.55 GPa and 3.8 GPa, respectively. Failure strains for fibers from the ASF and SPF/DB cycled panels were 0.011 and 0.013, respectively. The longitudinal modulus for fibers from the ASF and SPF/DB cycled panels were 330 GPa and 315 GPa, respectively. Thus, the SPF/DB cycle did not seem to affect the fiber properties to a large extent.

Typical fiber moduli for the SCS-6 fibers are 400 GPa [5], however, the moduli measured in this study were about 15 percent lower than this value. The gage lengths used in the present study were typically 75 mm. Since the crosshead displacement was used to calculate fiber strain, the accuracy of fiber strain and therefore modulus, increased with the gage length using the current test technique. This is due to a built-in fixed elongation in the load-train of the test machine. This fixed displacement becomes smaller compared to the total elongation of the fiber as the fiber gage length becomes longer. Figure 6 shows the variation of modulus with gage length for virgin fibers obtained from the manufacturer (Textron Specialty Materials). For a gage length of 300 mm the modulus was within 2 percent of the reported modulus of 400 GPa. For a gage length of 75 mm, the measured modulus for the virgin fibers was 337 GPa which is also within the scatter-band of the ASF fiber data. Using the current test technique, for a gage length of 75 mm, there is a 15 percent error in modulus. Nevertheless, the measurments do provide a means for comparing the moduli. Since the etched fibers gave essentially the same strength and modulus as the virgin fibers, it is assumed that the etching process did not affect the fibers.

Failure Mechanisms

The large differences in the UTS (Table 1) and fatigue strength (figure 4) of the ASF and SPF/DB cycled panels cannot be explained by the small differences in matrix and fiber properties. Also, the differences in the failure surfaces of the two specimen types (figure 5) cannot be explained by differences in the fiber and matrix properties. The present section explains how differences between the fiber/matrix interfaces can explain the observed differences in the failure surfaces and in the UTS (Table 1) and fatigue life (figure 4).

In the case of static tension loading, when a matrix crack encounters the fiber/matrix interface it can either break the fiber and move ahead or it can move ahead by debonding the interface for some distance. In the ASF specimens a crack that reaches a 0 deg fiber finds a path of less resistance along the weaker fiber/matrix interface and starts debonding it and running along the length of the fiber for some distance. This crack then exerts a much lower stress intensity at the fiber. This progression of damage along fiber/matrix interfaces dissipates more energy leading to the higher UTS values observed in the ASF specimens. In the SPF/DB cycled specimens, with the stronger interface, the matrix crack which reaches the interface does not debond the fiber/matrix interface, but instead breaks the fiber when the stress intensity at the crack tip reaches a critical value. This process of the matrix crack breaking through the fibers leads to lower UTS values.

The failure mechanisms postulated above are supported by the observations of the fracture surfaces in figure 5, in which the ASF specimens show an irregular fracture surface with protruding fiber ends and the SPF/DB cycled specimens show flush fiber ends. The evidence of split 90 degree fibers in the SPF/DB cycled specimens (figure 5) also suggests that the interface was sufficiently strong to transmit enough load into the fiber to cause it to split. It is assumed that the interface in the ASF was not this strong since no fibers were observed. The lower fatigue life in the SPF/DB cycled specimens are also explained by the same failure mechanism described above for static tension loading.

Fiber/matrix interface

The differences between the failure surfaces (figure 5) suggest that the fiber/matrix interface in the ASF and the SPF/DB panels is different. The present section describes the results of a microstructural analysis of the fiber/matrix interface in the ASF and SPF/DB cycled panels. The composite stress-strain response, which can be used to assess the strength of the fiber/matrix interface [5], is also examined in greater detail.

The fiber/matrix interfaces, as shown in figure 7, are significantly different between the ASF and SPF/DB cycled panels. The micrographs in figure 7 were obtained from oblique sections of the composite in order to exaggerate the fiber/matrix interface region. The specimens were etched with a solution of 100 ml H_2O, 6 ml HNO_3, and 3 ml HF. The ASF section (figure 7(a)) indicates a layered interface region as compared to the SPF/DB cycled section in which the titanium appears to have reacted with the carbon and the silicon, as a result of the high temperature SPF/DB cycle. Preliminary energy dispersive spectroscopy indicates the titanium reacts more extensively with the SiC at the interface during the SPF/DB process. This altered interface could be either stronger or weaker than the multi-layered interface region in the ASF panel. The interface layers in the ASF panel could have weaker interfaces in shear than that of the SPF/DB specimens.

Figure 8 shows a typical stress-strain response for the SPF/DB cycled specimens. The formation of the knee in the stress-strain curve at stress levels well below the yield stress of the matrix is typical of SCS-6/Ti-15-3 composites. Johnson et al [5] observed this knee for a range of composite layups and attributed the formation of the knee to the failure of the fiber/matrix interface in the off-axis plies. After the first loading-unloading cycle, a knee is observed on each subsequent cycle but at a lower stress level than the first cycle. As noted in reference 5, the knee on the first loading-unloading cycle corresponds to failure of the fiber/matrix interface while the subsequent knees are a result of the residual thermal stresses that developed during fabrication of the composite. The difference in the knee stress levels between the first and the subsequent cycles is, therefore, an indicator of the strength of the fiber/matrix interface in the off-axis plies. Note that the off-axis plies are being transversely loaded and this results in high tensile stresses in the radial (to the fiber) and tangential directions at the fiber/matrix interface. Therefore, the difference in the knees between the first and subsequent cycles is an indicator of the tensile strength of the interface.

The stress-strain response of the ASF specimens was similar to that of the SPF/DB specimens except that the average stress level at which the first knee in the stress-strain response was observed for the SPF/DB cycled specimens was about 50 percent greater than that for the ASF specimens (see Table 1). The difference between the first and subsequent knees for the SPF/DB specimens is about 25 percent greater than that in the ASF specimens. This suggests that the tensile strength of the fiber/matrix interface in the SPF/DB cycled specimens may be somewhat greater than that in the ASF specimens. The fact that the subsequent knee is much higher for the SPF/DB cycled specimens than the ASF specimens indicates that the residual stresses around the fiber may be greater. A simple finite element micromechanics analysis [7] indicates that the residual stresses will be 7 percent greater just because of the 9 percent higher modulus of the SPF/DB cycled titanium, thus supporting the experimental data.

Figure 9 shows close-up views of the same failure surfaces that were shown in figure 5. The 0 deg fibers are coming out of the paper. The region between the two 0 degree plies is a channel left behind by a 90 degree fiber that was pulled away. In the ASF specimen this channel appears to have a single uniform texture indicating a clean break at the fiber/matrix interface. The channel in the SPF/DB cycled specimen shows at least two different textures indicating that some interface layer was partly left behind. Thus, the interface layer was equally bonded to the fiber and matrix for the SPF/DB specimens but not for the ASF specimens. Hence, one of the interfaces in the ASF specimens, shown in figure 7, was relatively weak.

CONCLUDING REMARKS

The effects of a simulated superplastic forming/diffusion bonding (SPF/DB) temperature cycle on the tensile static and fatigue properties of a 3-ply [0/90/0] SCS-6/Ti-15V-3Al-3Cr-3Sn composite were investigated. The composite showed a 9 percent increase in modulus but a 25 percent decrease in UTS and a 30 percent decrease in failure strain as a result of the SPF/DB process. The fatigue endurance limit for the SPF/DB cycled specimens was 50 percent lower than the ASF specimens. The fracture surface of the ASF specimens was very irregular with significant fiber pull-out as compared to the planar fracture surface of the SPF/DB cycled specimens where fiber ends were flush with the surface.

The fibers leached from the ASF and SPF/DB cycled panels were only marginally different as a result of the high temperature cycle. The large changes in the UTS, the fatigue life, and the fracture surfaces due to the SPF/DB cycle are believed to be a result of changes in the strength of the fiber/matrix interface and the residual radial stresses around each fiber. A stronger interface in the SPF/DB cycled specimens leads to higher stress intensities at the fiber resulting in early fiber failures. The weaker interface in the ASF specimens results in cracks running along the length of the fibers and thereby dissipating more energy and lowering stress intensities at the fibers. This results in fiber failures at much higher loads. The fracture surfaces of the ASF and the SPF/DB cycled specimens are consistent with these different failure mechanisms.

A microstructural analysis of the fiber/matrix interface revealed a multi-layered interface in the ASF composite. The SPF/DB cycle altered the multi-layered interface. In both cases, the stress-strain response showed a knee typical of fiber/matrix interface failure. Differences between the first and subsequent knees for the SPF/DB cycled specimens and the ASF specimens suggested the SPF/DB cycle strengthened the fiber/matrix interface. Differences between appearances of the fracture surfaces were consistent with differences between interface strengths.

In short, for the laminates in question, the SPF/DB cycle apparently increased the matrix strength and modulus, had little or no effect on fiber strength and modulus, but significantly reduced both the static and fatigue strength of the laminate. This strength reduction is due to an elevation of residual radial compressive stresses and an apparently stronger interfacial bonding between the fiber and matrix.

REFERENCES

1. Harmon, D. M., Saff, C. R., and Sun, C. T., "Durability of Continuous Fiber Reinforced Metal Matrix Composites," AFWAL-TR-87-3060, October 1987.

2. Kaneko, R. S., and Woods, C. A., "Low-Temperature Forming of Beta Titanium Alloys," NASA Contractor Report 3706, September 1983.

3. Rosenberg, H. W., "Ti-15-3: A New Cold-Formable Sheet Titanium Alloy," Journal of Metals, Vol. 35, No. 11, November 1986, pp.30-34.

4. Wawner, F. E., Jr., "Boron and Silicon Carbide/ Carbon Fibers," Fibre Reinforcements for Composite Materials, Ed. A. R. Bunsell, 1988, Elsevier Science Publishers B. V., Amsterdam, The Netherlands.

5. Johnson, W. S., Lubowinski, S. J., Highsmith, A. L., Brewer, W. D., and Hoogstraten, C. A., "Mechanical Characterization of SCS_6/Ti-15-3 Metal Matrix Composites at Room Temperature," NASP Technical Memorandum 1014, April 1988.

6. Johnson, W. S., "Fatigue Testing and Damage Development in Continuous Fiber Reinforced Metal Matrix Composites," Metal Matrix Composites: Testing, Analysis and Failure Modes, ASTM STP 1032, W. S. Johnson, Ed., 1989.

7. Highsmith, A. L., Shin D., and Naik, R. A., "Local Stresses in Metal Matrix Composites Subjected to Thermal and Mechanical Loading," Metal Matrix Composites: Testing, Analysis and Failure Modes, ASTM STP 1032, W. S. Johnson, Ed., 1989.

Table 1.- Static properties of ASF and SPF/DB [0/90/0] composite panels.

	Modulus (GPa)	Poisson's Ratio	Knee (MPA) First	Subsequent	UTS (MPa)	Failure Strain
ASF panel	156	0.21	67±2	46±3	1070	0.0105
SPF/DB panel	170	0.33	103±3	77±2	805	0.0073

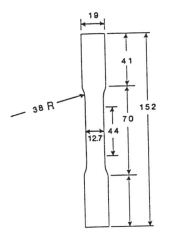

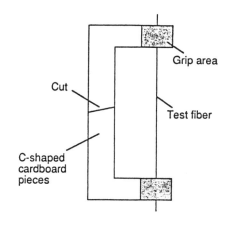

Fig. 1. Dog-bone specimen.

All dimensions in mm.

Fig. 2. C-shaped fixture for fiber
testing.

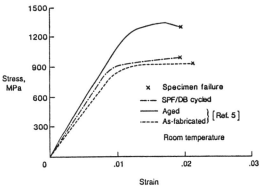

Fig. 3. Unreinforced Ti-15-3 stress-strain curves.

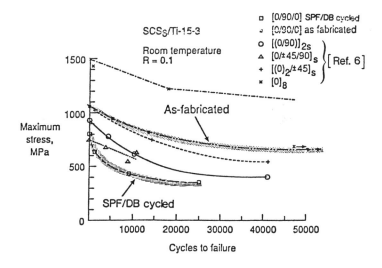

Fig. 4. Fatigue data for SCS-6/Ti-15-3 composites.

101

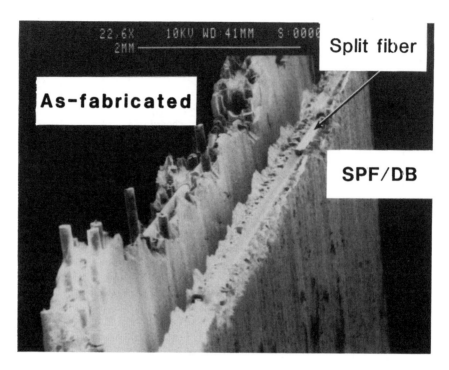

Fig. 5. Fracture surface morphology for statically tested specimens.

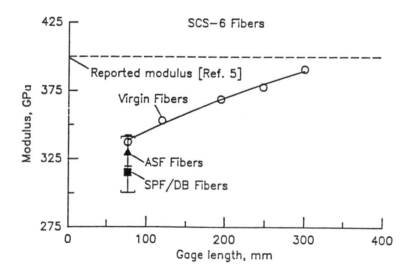

Fig. 6. Fiber modulus variation with gage length.

ASF   SPF/DB cycled

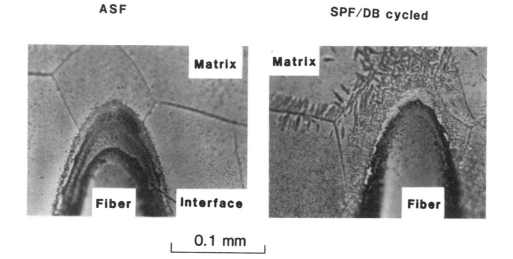

|⎣ 0.1 mm ⎦|

Fig. 7. Micrographs of the fiber/matrix interface region.

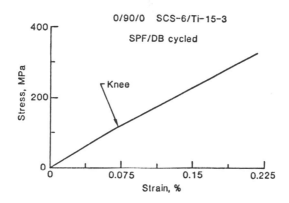

Fig. 8. Typical stress-strain curve for SCS-6/Ti-15-3 composite.

ASF   SPF/DB cycled

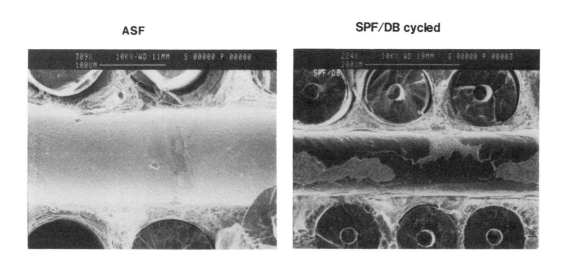

Fig. 9. Micrograph showing channel left behind by a pulled out fiber.

Elastic Crack Bridging in Ceramic and Intermetallic Matrix Composites

L. R. DHARANI AND L. CHAI

Department of Mechanical and Aerospace Engineering and Engineering Mechanics
University of Missouri-Rolla
Rolla, Missouri 65401-0249

ABSTRACT

A micromechanics analytical model based on the consistent shear lag theory is developed for predicting the failure modes in a fiber reinforced unidirectional ceramic and intermetallic matrix composite. The model accounts for the relatively large matrix stiffness. The fiber and matrix stresses are established as functions of the applied stress, crack geometry, and most importantly, the microstructrural properties of the constituents. From the predicted stress, the mode of failure is established based on the point stress criterion.

INTRODUCTION

Observations of failure in very brittle matrix composites have led to a strong interest by contemporary researchers in the so-called steady-state cracking condition. This condition is manifested in a long matrix crack extending from the free edge of a unidirectional composite. The crack is perpendicular to the fiber (and loading) direction and bridged by the fibers. The first example of modeling this failure mode is represented in the classical works of Aveston, Cooper and Kelly [1], Aveston and Kelly [2]. A More recent and pertinent analytical work on this and related problems is by Budiansky, Hutchinson and Evans [3], Marshall, Cox and Evans [4], Budiansky and Amazigo [5] and Dharani, Chai and Pagano [6]. A brittle matrix reinforced by strong and relatively ductile fibers with a weak interface generally leads to steady-state cracking discussed above. The process is insensitive to notch length. A non-steady state cracking in which failure is notch sensitive may result if the matrix is brittle but the other two conditions for the steady-state cracking are not met.

Recently, a micromechanics analytical model was developed by Chai and Dharani [7] for characterizing the non-steady state cracking in brittle matrix composites. A unidirectional composite subjected to uniaxial tensile load parallel to the axis of the fibers was considered for the analysis with two (finite length) cracking configurations. In the first configuration the matrix crack has passed through the fibers and is characterized by a strong interface and weak fibers. Whereas in the second configuration the matrix crack is bridged by intact fibers and is characterized by strong fibers and strong interface. The analysis also included longitudinal debonding at the main crack tips of the two cracking configurations. The model is based on a consistent shear lag theory [8] and accounts for a relatively large matrix stiffness. In this paper, we extend the above model to include another non-steady state cracking configuration which is essentially a combination the two configurations considered earlier. An initial crack in the form of broken fiber and matrix phases is assumed to extend in a crack bridging mode in which the bridging phase is relatively ductile as compared to the fractured phase. This failure mode is illustrated in Fig. 1 (b). As before the fiber-matrix interface is moderately strong so that slipping at the interface does not take place and it can be considered as elastic. For this configuration we present shear lag stresses and displacements in the fiber and the matrix as functions of the applied stress, crack geometry, and the microstrctrural properties of the constituents. From the predicted stresses, the mode of failure is studied based on a point stress criterion.

ANALYTICAL MODEL

A unidirectional composite subjected to a uniaxial tensile load parallel to the axis of the fibers is considered in the analysis. Using the superposition principle [7], the problem can be considered as the superposition of two problems. The first problem consists of a plate without a crack acted by the same applied stress as the original one at infinity. The other one is that of a plate with a crack acted by the same applied stress at the crack faces instead of at infinity. Because of the symmetry only the upper half of the lamina is considered. Unlike the classical shear lag models the present model takes into account the high stiffness of the matrix phase by including transverse and axial stresses in the matrix. Therefore, both the axial and transverse equilibrium equations have to be considered in the analysis. By proposing a set of approximate stress-displacement relations the equilibrium equations can be written in terms of axial and transverse displacements. The approximate stress-displacement relations are [7]:

$$\sigma_{nr}^x = A_{11} \frac{u_{n2} - u_{n1}}{d/2} + A_{12}(D_{vf}v_{n1,y} + D_{vm}v_{n2,y})$$

$$\sigma_{nl}^x = A_{11} \frac{u_{n1} - u_{(n-1)2}}{d/2} + A_{12}(D_{vf}v_{n1,y} + D_{vm}v_{(n-1)2,y})$$

$$\tau_{nr} = A_{66}\left[D_{uf}u_{n1,y} + D_{um}u_{n2,y} + \frac{v_{n2} - v_{n1}}{d/2} \right]$$

$$\tau_{nl} = A_{66}\left[D_{uf}u_{n1,y} + D_{um}u_{(n-1)2,y} + \frac{v_{n1} - v_{(n-1)2}}{d/2} \right]$$

$$\sigma_{nf}^y = C_{12}D_{um} \frac{u_{n2} - u_{(n-1)2}}{b} + C_{11}v_{n1,y}$$

$$\sigma_{nm}^y = B_{12}D_{uf} \frac{u_{(n+1)1} - u_{n1}}{b} + B_{11}v_{n2,y}$$

$$\tau_{nf} = A_{66} \frac{v_{n2} - v_{(n-1)2}}{d} + A_{66}u_{n1,y}$$

$$\tau_{nf} = A_{66}\, \frac{v_{(n+1)1} - v_{n1}}{d} + A_{66}u_{n2,y} \tag{1}$$

where σ_n and τ_n are the normal and shear stresses in an element n, subscripts f and m stand for the fiber and the matrix, l and r stand for the left and the right sides of the fiber element, d is the fiber spacing, u_{n1} and v_{n1} are the transverse (x) and axial (y) displacements of the fiber n and u_{n2} and v_{n2} are the corresponding matrix displacements, and A_{ij}, B_{ij}, and D_{ab} are the stiffness parameters which are functions of material and geometric constants. By defining a normalized longitudinal coordinate z = y/d, the equilibrium equations for a typical element (n) can be written in terms of displacements and material constants as:

$$2A_{11}(u_{(n-1)2} - 2u_{n1} + u_{n2}) - (A_{12}D_{vm} + A_{66}C_f)(v_{(n-1)2,z} - v_{n2,z}) + A_{66}C_f u_{n1,zz} = 0$$

$$2A_{66}(v_{(n-1)2} - 2v_{n1} + v_{n2}) - (A_{66} + C_{12})D_{um}(u_{(n-1)2,z} - u_{n2,z}) + C_{11}C_f v_{n1,zz} = 0$$

$$2A_{11}(u_{n1} - 2u_{n2} + u_{(n+1)1}) - (A_{12}D_{vf} + A_{66}C_m)(v_{n1,z} - v_{(n+1)1,z}) + A_{66}C_m u_{n2,zz} = 0$$

$$2A_{66}(v_{n1} - 2v_{n2} + v_{(n+1)1}) - (A_{66} + B_{12})D_{uf}(u_{n1,z} - u_{(n+1)1,z}) + B_{11}C_m v_{n2,zz} = 0 \tag{2}$$

The equilibrium equations for the free edge elements are similarly obtained by substituting the corresponding boundary conditions into those for the typical elements. The final equilibrium equations can be written in a matrix form as:

$$[M]\{u_{,zz}\} + [C]\{u_{,z}\} + [K]\{u\} = \{0\} \tag{3}$$

where:

$$\{u\} = (......\ ,\ u_{n1},\ v_{n1},\ u_{n2},\ v_{n2},\ ......)^T_{2N}$$

is the basic unknown vector, and $[M]$, $[C]$ and $[K]$ are the coefficient matrices ($2N \times 2N$) corresponding to the second, first and zeroth order derivatives of the basic unknown vector respectively [7] and z is the normalized coordinate y/d. By assuming the solutions as exponential functions of the normalized longitudinal coordinates, the general solution of the differential equations is reduced to the classical eigenvalue-eigenvector problem. Each pair of eigenvalues and eigenvectors represents a component of the displacement. The general solution with 4N constants is given by:

$$\{r\} = \sum_{K=1}^{M} [C_{2K-1}(\{P_K\}\cos b_K z - \{Q_K\}\sin b_K z)e^{a_K z} + C_{2K}(\{Q_K\}\cos b_K z$$

$$+ \{P_K\}\sin b_K z)e^{a_K z}] + \sum_{K=2M+1}^{4N} C_K\{L_K\}e^{\lambda z} \tag{4}$$

where, M is the number of the pairs of the conjugate complex eigenvalues, a_K and b_K are the real and imaginary parts of the Kth complex eigenvalue, $\{P_K\}$ and $\{Q_K\}$ are the real and imaginary parts of the Kth complex eigenvetor respectively and C_K is the integral constant corresponding to the Kth eigenvalue and eigenvector and can be determined by satisfying the boundary and continuity conditions appropriate to the damage configuration.

RESULTS AND DISCUSSION

We present some typical results in terms of stresses and crack opening displacements for the two cracking configurations shown in Fig. 1 (a) and 1 (b). The properties used in this analysis are: fiber modulus, $E_f = 200 GPa$, matrix modulus, $E_m = 85 GPa$, fiber size, $b = 8\mu m$, fiber volume fraction, $c_f = 0.5$. An initial crack is formed by a continuous fracture of both the matrix and the fiber. Then this initial unbridged crack is assumed to extend in a bridged fashion. The material of the bridging ligament is relatively more ductile than that of the fractured ligament. As a result, we will present at two crack bridging cases; in the first case fibers are bridging a matrix crack and in the second a ductile matrix is intact and connects the broken fibers. For convenience, the stresses are normalized as follows:

$$\overline{\sigma}_f = \sigma_y^f \frac{E_m}{d}$$

$$\overline{\sigma}_m = \sigma_y^m \frac{E_m}{d}$$

where σ_y^f and σ_y^m are the axial stresses in the fiber and the matrix respectively, E_m is the elastic modulus of the matrix and d is the fiber spacing. In the following discussion, for the sake of brevity, we will refer to the configuration shown in Fig. 1 (a) as continuous cracking mode (ccm) and the other shown in Fig. 1 (b) as cracking bridging mode (cbm).

In all cases presented here, a unit remote stress is applied on the matrix phase so that the fiber remote stress is scaled in the ratio of E_f/E_m . The maximum stresses in the fiber and the matrix along the original crack plane are computed and plotted in Figs. 2 and 3 for ductile and brittle fibers respectively. In Fig. 2 the solid lines correspond to ccm and the dotted lines correspond to cbm. For both modes of crack growth the maximum stresses in the fiber and the matrix increase as the main crack length increases. However, for the crack bridging mode, at a given crack length, matrix stress decreases and the fiber stress increases as compared to continuous cracking mode. This indicates if the matrix crack is bridged elastically by ductile fiber, the fiber is more likely to fracture. Figure 3 shows the corresponding results for a brittle fiber so that the crack bridging ligaments are of matrix. The difference in the matrix stress at a given crack length between cbm and ccm is very significant. If the failure strain of the matrix is large then the crack bridging ligament can elongate more freely thereby providing higher toughness. Figure 4 gives the crack face displacement for various normalized lengths of the main crack. As the crack length increases the crack opening at a fixed distance from the crack tip increases indicating that the steady state cracking condition has not been reached.

We have drawn some qualitative conclusions based on the point stress failure criterion. Because of the inherent singularity at the crack tip the point stress criterion would give vastly different results depending on the selection of the point where the criterion is applied. A more rigorous energy release rate and stress intensity approaches are required in order to draw any quantitative conclusion and this work is under progress by the authors.

ACKNOWLEDGEMENTS

This work was initially supported by National Aeronautics and Space Administration through a contract to Incubator Technologies, Inc., Rolla, Missouri, under the SBIR program. The authors are grateful to Dr. Christos C. Chamis, NASA Lewis-Research Center, for his help and suggestions during the course of the work.

REFERENCES

1. Aveston, J., C. A. Cooper, and A. Kelly, "Single and Multiple Fracture," in The Properties of Fiber Composites, Conference Proceedings, NPL, 1971, pp. 16-26.

2. Aveston, J., and A. Kelly, "Theory of Multiple Fracture of fibrous composites," J. Mater. Sci. 8, 1973, pp. 352-362.

3. Budiansky, B., J. W. Hutchinson, and A. G. Evans, "Matrix Fracture in Fiber-Reinforced Ceramics," J. Mech. Phys. Solids, 34(2), 1986, pp. 167-189.

4. Marshall, D. B., B. N. Cox, and A. G. Evans, "The Mechanics of Matrix Cracking in Brittle-Matrix Fiber Composites," Acta Metall., 33(11), 1985, pp. 2013-2021.

5. Budiansky, B. and J. C. Amazigo, "Toughening by Aligned, Frictionally Constrained Fibers," J. Mech. Phys. Solids, 37(1), 1989, pp. 93-109.

6. Dharani, L. R., L. Chai and N. J. Pagano, "Steady-State Cracking in Ceramic Matrix Composites," Composite Sci. Tech. 1989 (in review)

7. Chai, L. and L. R. Dharani, "A Micromechanics Model for Prediction of Failure Modes in Ceramic Matrix Composites," Final Report NASA Contract No NA53-25333, August 1988.

8. Sendeckyj, G. P. and W. F. Jones, "Improved First Order Shear Lag Theory for Unidirectional Composites with Broken fibers," Engng. Fracture Mech., (to appear)

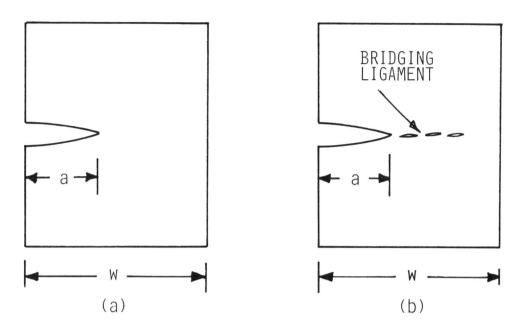

Fig. 1 Cracking configurations: (a) continuous cracking mode (ccm)
(b) crack bridging mode (cbm).

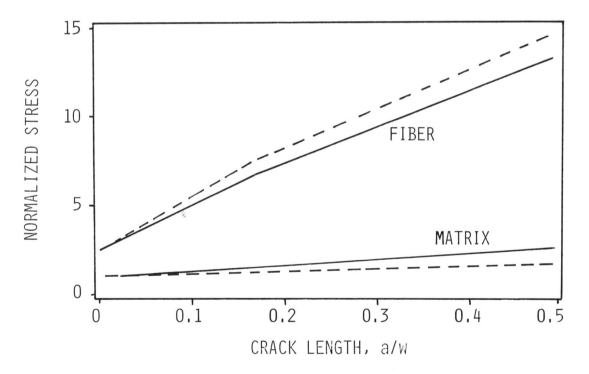

Fig. 2 Maximum fiber and matrix stresses for a ductile fiber composite
(— ccm, --- cbm).

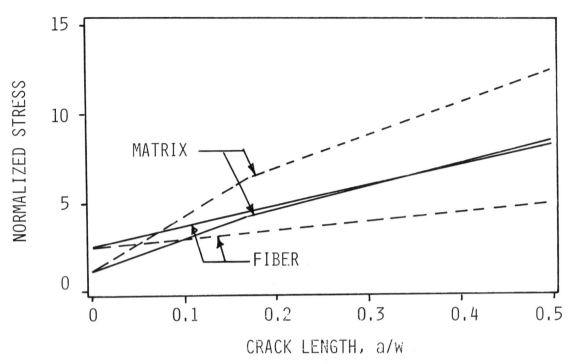

Fig. 3 Maximum fiber and matrix stresses for a brittle fiber composite
(— ccm, --- cbm).

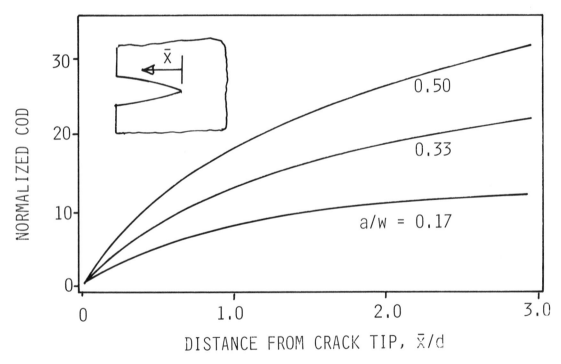

Fig. 4 Crack opening for a ductile fiber composite.

Whisker Coating in Whisker Reinforced Al Matrix Composites

CHI HWAN LEE, JOON HO SEO AND **SANG HOON HAN**
Department of Metallurgical Engineering
College of Engineering
Inha University
Incheon, KOREA

ABSTRACT

Effects of nickel coating for whiskers have been investigated on tensile property and morphology. Whisker coating was carried out by using electroless deposition method. The potassium tatanite ($K_2O.6TiO_2$) whisker reinforced aluminum matrix composites were fabricated by the powder metallurgy (P/M) process.

After fabrication the room temperature tensile strength of the specimen with coated whisker appeared to be superior to that of the specimen with the uncoated whisker up to the whisker volume content of 30 percent. After thermal exposure it was observed that the room temperature tensile strength of the uncoated whisker specimen conspicuously decreases, whereas tensile strength of the specimen with coated whisker does not decrease even after 50 hours of exposure at 500°C. Scanning electron micrography observation on whisker morphology of these specimens showed that the uncoated whisker has a very rough surface, whereas the coated whisker remains unchanged. Nickel coating on the potassium tatanite whisker was found to be effective as a diffusion barrier for chemical reaction and improvement of wetting between the whisker and Al matrix during fabrication and thermal exposure.

INTRODUCTION

It is well known that whisker reinforced metal matrix composites exhibit desirable characteristics such as high specific strength and stiffness, high temperature capability, and high wear resistance (1-3). SiC whisker reinforced Al alloy composites have many interesting properties which are attractive for aircraft structural application. However this material also has some limitations, such as particularly low tensile ductilities, low fracture toughness, etc.

In addition the machining is very difficult because of the presence of the extremely hard and abrasive ceramic SiC whiskers in a soft matrix. Compared with SiC whisker reinforced Al alloy composites, potassium tatanite whisker reinforced Al composites can be machined easily with ordinary tools because potassium tatanite whisker is not as hard as SiC whisker. Furthermore, potassium tatanite whisker has a potential for application in commercial fields because of the very cheap price. Previous work (4) on this whisker indicated that during the fabrication of potassium tatanite whisker reinforced Al composites by squeeze casting, a chemical reaction between the whisker and molten Al makes it difficult to improve the mechanical property of the composite. Therefore it is necessary to fabricate the composites with a moderate volume fraction of whisker. Using the P/M method, it is possible to fabricate composites at temperatures lower than the melting point at which the chemical reaction becomes active. It has been known that the P/M method is limited to the excellent wetting between the whisker and Al matrix because of the solid state fabrication condition. Therefore the coating method for whiskers is the most promising one to ensure the elimination of chemical reaction and improvement of wetting during fabrication. A very limited number of works on coating of potassium tatanite whiskers can be found in the literature.

The purpose of this paper is to examine the effects of nickel coating for the whisker on tensile properties and morphology of potassium tatanite whisker reinforced aluminum composites.

EXPERIMENTAL PROCEDURE

Potassium tatanite whisker was used as the reinforcement in this study, and its properties are listed in Table 1. A commercial grade of pure Al powder containing 98 wt. percent of -325 mesh size was used as the matrix. Electroless deposition solution (Japan Kanigen Co.) was used for the whisker coating solution. The coating process is the same as reported in a previous paper (5,6). Ultrasonic stirring apparatus in a cylindrical container containing solution with the mixture of Al powder and whisker was provided in order to avoid whisker entanglement and the homogeneous distribution of whisker and uniformity of coating layer thickness. Deposition time taken was about 30 seconds in order to obtain a thin coating layer under the deposition condition as described in Table 2. Coating thickness obtained was about 0.05 μm. The nickel coating layer appeared to contain several percent of dissolved phosphorous in solution during processing. Nickel coated whisker was mixed with Al powder in a sodium oleate solution by the use of ultrasonic apparatus and then dried. Subsequently the mixture shown in Figure 1 was cold pressed and then compacted by hot pressing under a pressure of 100 MPa in an argon atmosphere. Figure 2 illustrates the P/M process flow to produce specimens. The five rectangular plates (80 mm x 32 mm x 13 mm) fabricated have whisker volume fractions of 0, 10, 20, 30, and 40 percent, respectively. In addition specimens with uncoated whiskers were made for comparison. After fabrication, plates (coated and uncoated) were exposed at 500°C for 50 hours. Tensile tests were carried out in the temperature range of room temperature to 400°C. The specimens (80 mm x 15 mm x 3 mm) of unreinforced Al, whisker/Al composites, and Ni-coated whisker/Al composites were loaded using a testing machine equipped with an electrical furnace. Observations for microstructures and fracture surfaces of specimens were carried out by means of optical microscopy and SEM.

RESULTS AND DISCUSSION

Microstructures of whisker distribution in specimens with various Al powder sizes are presented in Figures 3(a) and (b). The whiskers appear to be black in contrast to the white part of the Al matrix. Whisker distribution in the Al matrix containing 98 percent of -325 mesh (fine powder) appears to be homogeneous as shown in Figure 3(a). However whisker distribution in the Al matrix containing 75 percent of -325 mesh (coarse powder) was found to be inhomogeneous because entangled whisker rich region (black part) is surrounded by large Al matrix grain about 50 μm in size as shown in Figure 3(b). The microstructural inhomogeneity is attributable to the scatter and the drop in mechanical properties as reported in Reference 7. Figure 4 illustrates stress-strain curves of room temperature tensile strength in specimens with different Al powder sizes. This plot shows that values of strength and fracture strain of specimens with coarse powder are lower compared to those of specimens with fine powder. SEM observation of fracture surfaces as shown in Figure 5 shows that specimens with fine powder reveal a dimple pattern characterizing ductile fracture, whereas specimens with coarse powder reveal a brittle fracture mode in the entangled whisker rich region. In particular, specimens with coarse powder appeared to fail in a brittle manner, resulting from cracks that propagate through the brittle whisker rich region under tensile stress. Therefore it is considered that whisker distribution in the Al matrix depends on matrix particle size. Figure 6 shows room temperature tensile strength of specimens with Ni-coated and uncoated whiskers. Tensile strength increases as the whisker volume content increases up to 20 percent. The increase in tensile strength was probably caused by closer packing of the reinforcement and smaller interwhisker spacing in the matrix as presented in previous work (8). However it is observed that as the whisker content reaches 30 to 40 volume percent, the value of the tensile strength does not increase at all. Thus the drastic downfall of the reinforcing effect may be attributed to the poor wetting between the whisker and the Al matrix at the interface. It is noted from Figure 7 that the density of the specimen decreases with increasing volume percent of the whisker. Especially, the specimen with 40 percent of whisker volume shows a relatively low value of 97.5 percent of theoretical density. The decrease in density due to increasing volume percent of the whisker results in an increase of microvoid formed by poor bonding between the whisker and the Al matrix. Consequently microvoid as a stress riser is detrimental to tensile stress and ductility in the whisker reinforced Al matrix composites. Therefore it is considered that mechanical property in the whisker reinforced Al composites largely depends upon the degree of wetting at the whisker-Al matrix interface.

In this study electroless Ni coating on the whisker surface was attempted in order to improve wettability and control chemical reaction at the potassium tatanite whisker and the Al matrix. It is shown that the room temperature strength of the specimen with the Ni-coated whisker is superior to that of the specimen with the uncoated whisker as shown in Figure 8. This would be caused by the improvement of wettability accompanied with interdiffusion between Ni-coating layer and the Al matrix during fabrication at 640°C. The Ni-coating effect on chemical reaction in the specimen exposed at 500°C for 50 hours is presented in Figure 9. From this figure, room temperature tensile strength of the specimen with the Ni-coated whisker appeared to be superior to that of the specimen with the uncoated whisker. After etching Al matrix with NaOH solution, results of SEM observations for the whisker surface are presented in Figure 10. Whisker surface of the specimen with the uncoated whisker seems to be considerably rougher than that of the specimen with the Ni-coated whisker. A rough layer of the whisker surface may be considered to be the chemical reaction zone formed between the whisker and the Al matrix during thermal exposure. The role of Ni-coating layer on the whisker appeared to be the diffusion barrier for chemical reaction between potassium tatanite whisker and the Al matrix and improved wetting at the interface. Figure 11 shows the tensile strength of the specimen at elevated temperatures. Tensile strength in all specimens tends to decrease as test temperature increases. The strength of the specimen with the whisker appears to be considerably superior to that of the unreinforced Al matrix at 400°C. Especially, the tensile strength of specimens with 30 volume percent was 180 MPa at 200°C and 130 MPa at 400°C. From the above results, it is concluded that the strength of potassium tatanite whisker/Al composites is significantly improved at elevated temperature by the addition of the whisker.

CONCLUSIONS

This investigation led to the following conclusions.

1.   Uniform distribution of potassium tatanite whisker in Al matrix was found to be dependent upon Al powder size.

2.   The room temperature tensile strength of the specimen with coated whisker appeared to be superior to that of the specimen with the uncoated whisker up to 30 percent of the whisker volume content.

3.   After thermal exposure, it was observed that room temperature tensile strength of the uncoated whisker specimen conspicuously decreases, whereas tensile strength of the specimen with coated whisker does not decrease even after 50 hours exposure at 500°C.

REFERENCES

1.   D. Webster, Metal Trans., 13A, 1982, p. 1151.

2.   D. F. Hasson, S. M. Hoover, J. of Material Sci., 20, 1985, p. 74.

3.   A. P. Divecha et al., J. Metals, 33, 1981, p. 21.

4.   H. Fukunaga et al., Transaction of the Japan Institute of Metals, Vol. 24, No. 9, 1983, p. 642.

5.   G. G. Gawrilov, "Electroless Nickel Plating," Redhill, England, Portcullis Press, 1979, p. 36.

6.   G. Gutzeit, "Plating and Surface Finishing," Vol. 46 (No. 10), 1959, p. 1158.

7.   T. G. Nieh, R. F. Karlak, J. of Materials Science Letters 2, 1983, p. 119.

8.   D. L. McDanels, Metall. Trans., Vol. 16A, 1985, p. 1105.

Table 1   Properties of Potassium Tatanite Whisker

Average Length	20-30 μm
Average Radius	0.3-0.5 μm
Density	3.58(g/cm³)
Tensile Strength	700 kg/mm²
Elastic Modulus	2.8 x 10⁴ kg/mm
Melting Point	1300-1350°C
Thermal Expasion Coefficient	6.8x 10⁻⁶/°C

Table 2   Electroless Nickel Plating Solution and Operating Condition

Composition		Operating Condition
Nickel Chloride	30 g/L	PH: 4-6
Sodium Hypophosphite	10g/L	Temperature: 363 K
Sodium Hydrooxyacetate	10g/L	Plating Rate: 13μm/h

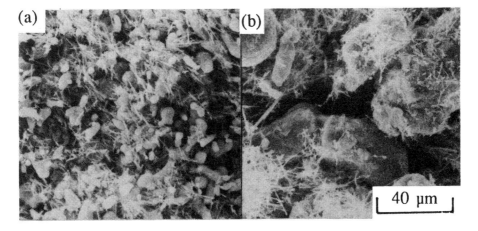

Fig.1   Appearences of $K_2O.6TiO_2$ whisker preform mixed with Al
powder of (a) 98 % of -325 mesh,(b) 75 % of -325 mesh.

Fig.2 Process flowchart for P/M fabrication of specimen.

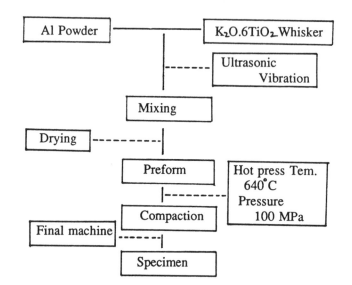

Fig.3 Microstructures of $K_2O.6TiO_2$/Al composites with Al powder of (a) 98 % of -325 mesh,(b) 75 % of -325 mesh.

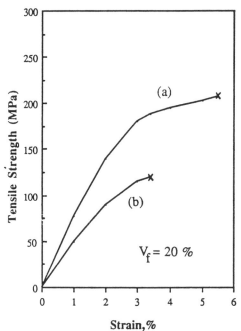

Fig. 4 Stress-Strain curves for $K_2O.6TiO_2/Al$ composites with (a) 98% of -325 mesh, (b) 75 % of -325 mesh.

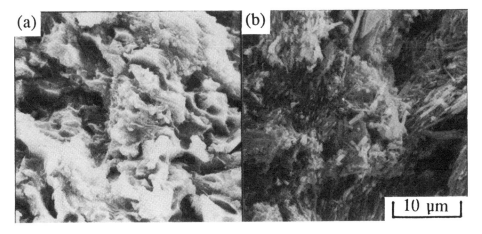

Fig.5 Fractograpy of $K_2O.6TiO_2/Al$ composites with different Al powder. (a) 98 % of -325 mesh, (b) 75 % of -325 mesh

117

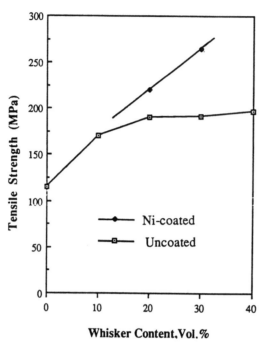

Fig. 6 Room temperature tensile strength VS. vol. fraction
in specimens.

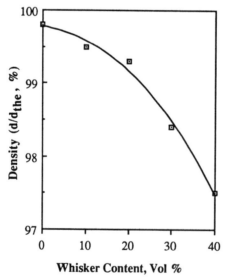

Fig. 7 Variation of density with whisker vol. fraction
in specimens.

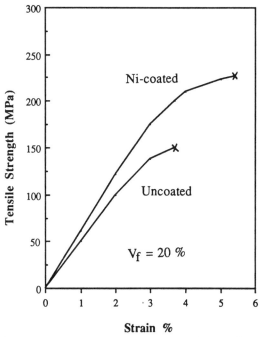

Fig.8   Stress-Strain curves for specimens thermal-exposed at 500°C for 50 hours.

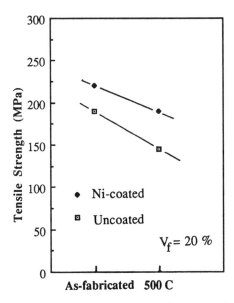

Fig. 9   Room temperature tensile strength of specimens as-fabricated and thermal exposed at 500°C.

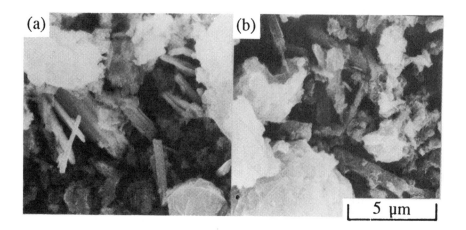

Fig.10 Whisker morphology of specimens exposed at 500°C for 50 h.
   (a) Ni-coated ,(b) Uncoated

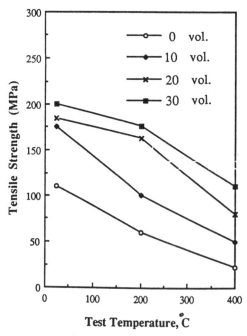

Fig.11 Tensile strength of $K_2O.6TiO_2$ whisker/Al composites
   and unreinforced Al at elevated temperatures.

120

Isothermal Life Prediction of Composite Lamina Using a Damage Mechanics Approach

NADER MOHAMED ABUELFOUTOUH
National Research Council Fellow
NASA Lewis Research Center

M. J. VERRILLI AND G. R. HALFORD
NASA Lewis Research Center
Cleveland, Ohio 44135

ABSTRACT

A method for predicting isothermal plastic strain fatigue life of a composite lamina is presented in which both fibers and matrix are isotropic materials. in general the fatigue resistance of the matrix, fibers, and interfacial material is needed to predict composite fatigue life. The composite fatigue life in this paper is predicted using only the matrix fatigue resistance due to inelasticity micromechanisms. The effect of the fiber orientation relative to loading direction is accounted for while predicting composite life. The application is currently limited to isothermal cases where the internal thermal stresses that might arise from thermal strain mismatch between fibers and matrix are negligible. The theory is formulated to predict the fatigue life of a composite lamina under either load or strain control. It is applied currently to predict the life of a tungsten-copper composite lamina at 260 C under tension-tension load control. The predicted life of the lamina is in good agreement with available composite low cycle fatigue data.

KEYWORDS

Fatigue life; Tungsten-copper; Metal matrix composite ; High temperature; Constitutive relationship; Life prediction

INTRODUCTION

Fatigue failure of a metal matrix composite (MMC) is a complex process. Failure modes can depend on the applied load, the properties of the matrix, fibers and interface, fiber volume fraction and orientation, and on the service temperature and environment. Depending on these factors, the active failure mode can be matrix dominated, fiber dominated, fiber/matrix interfacial failure, or self-similar fatigue damage [1]. A fatigue life prediction method has to consider the most active mode of failure to obtain a good estimate for life.

In this work a tungsten reinforced copper composite

121

containing unidirectional continuous fibers has been studied .
For specimens whose fibers were oriented parallel to the load
axis, the failure mode was found to be matrix dominated.
Therefore for the ensuing analysis this mode of failure was
assumed to dominate for all angles of fiber orientations.

It is assumed for simplicity reasons alone that the
composite fatigue failure can be considered as a sequence of
two events where complete fatigue failure of the matrix is
followed by immediate complete failure of the fibers. Matrix
cracking introduces additional cyclic axial and shear stresses
to the fibers. The localized nature of these stresses severely
reduces the fiber residual life and therefore its contribution
to composite fatigue life can be neglected.

The analysis in this paper is limited to composites with a
strong fiber-to-matrix bond and no interfacial phase, Fig. (1).
In tungsten-copper composites the bonding is excellent and the
two constituents are mutually insoluble in each other.

CONSTITUTIVE RELATIONSHIP

The state variable constitutive relationship used in this paper
is derived in [2] and is represented by the following two
equations for the inelastic strain rate and the rate of
evolution of the state variable ,X, referred to herein as
"resistance to flow",

$$\dot{e}_{ij}^{i} = A0 \; (J2/X^{2})^{r} \; s_{ij} \tag{1}$$

$$\dot{X} = F(X) \; J2^{r+1} = \pm \exp(CX+D) \; J2^{r+1} \tag{2}$$

where e_{ij}^{i} is the inelastic strain tensor, $A0$, C , D and r are
material constants. Values of the constants at 260 C are given
in Table 1. J2 is the second invariant of the deviatoric
stress tensor s_{ij}. The sign of the rate of evolution is
positive if the material hardens and negative if it softens or
is damaged. The superscript i refers to inelastic strain and
the dot represents the first derivative with respect to time.
The function F(X) is a material state function that controls
response and determines life. The state variable increases for
hardening materials and then achieves a stabilized or "shake
down" value, X_s, after which the material resistance
decreases until catastrophic failure occurs, Fig. (2a). It
follows for the continuously applied cycles that the
integration of $J2^{r+1}$ over the applied cycle increases with the
increase of X_s and decreases with its decrease. This integration is called the loading function L and is represented in
Fig. (2b) as a function of the number of applied cycles. The
value of this integration is assumed constant and equal to its
value at shake down L_s. Using this assumption and integrating
Eq. (2) over the entire life, it follows that the relationship
between the number of cycles to failure, N_f, under continuously
applied cycles is related to the initial and shake down

resistances as follows

$$N_f \, L_s = - [\, 2 \exp(-CX_s-D) - \exp(-CX0-D)-\exp(-D)]/C \quad (3)$$

where the resistance at failure equals zero. Eq. (3) is used
to predict fatigue life whenever L_s and X_s are known.

SHAKE DOWN RESISTANCE

The evaluation of the shake down resistance X_s of the composite
constituents requires an experimental relationship between
this resistance and the applied stress under load control or
the applied strain under strain control. The life prediction
method used in this paper requires the evaluation of the shake
down resistance X_s and load function L_s. These values are
employed in Eq. (3) to calculate the number of cycles to failure. The shake down resistance is calculated iteratively. A
value for the shake down resistance is first assumed and used
in the constitutive relationship to calculate the strain or
stress response under load or strain control, respectively. The
calculated response is applied to the experimental relationship
between the resistance X_s and the applied stress or strain to
update X_s. The updated resistance is then used to update the
calculated response. This process continues until change in
the calculated shake down resistance over an iteration is
negligible. The load function for a constituent is calculated
from its constitutive relations taking the resistance constant and equal to its value at shake down.

LOCAL STRESSES AND STRAINS IN COMPOSITE LAMINA

Strain Control

In this section the local stresses and strains in the
constituents of a composite lamina with perfectly bonded
continuous fibers are considered. An element of the composite
lamina is shown in Fig. (1). For convenience this element is
assumed to be composed of uniformly spaced square fibers and
matrix strips and is subjected to in-plane stress which is
produced by a biaxial strain control. It is required to
evaluate the stress and strain at any point in the lamina. The
evaluation begins by introducing a continuum displacement field
which represents the longitudinal displacements along the
centerline of fibers and matrix strips. The continuum
longitudinal strain obtained by differentiating the
displacement field is assumed equal to that of fibers and
matrix at any time. Therefore the continuum longitudinal
strain rate equals the fiber and matrix longitudinal strain
rate. The fiber and matrix shear and strain in the lateral
direction of the fibers are not equal to the continuum strain
obtained by differentiating the displacement field. The
equality of such strains contradicts the equilibrium at fiber
surface due to the difference between fiber and matrix
mechanical properties. The shear and lateral continuum strain
are smaller than those of the matrix and larger than those of

fibers. Let us assume a linear displacement field, and consider the problem shown in Fig. (1) where the composite is subjected to biaxial strain rates. It can be shown that the continuum longitudinal, lateral and shear strain rates in axes parallel and perpendicular to the fibers are given by

$$\dot{e}_{11c} = \dot{e}_x \cos^2\theta + \dot{e}_y \sin^2\theta$$
$$\dot{e}_{22c} = \dot{e}_x \sin^2\theta + \dot{e}_y \cos^2\theta \qquad\qquad (4)$$
$$\dot{e}_{12c} = -(\dot{e}_x - \dot{e}_y) \sin\theta \cos\theta$$

where $\dot{e}_x$ and $\dot{e}_y$ are the applied strain rates in the **x** and **y** directions respectively. The continuum strains e_{11c}, e_{22c} and e_{12c} are measured along the fiber direction, perpendicular to the fiber direction and in-plane with these two directions, respectively. θ is the angle between the **x** axis and fiber direction. The matrix lateral and shear strain rates are obtained from the continuum strain rates by assuming that the displacement along the centerlines of the fibers and the matrix strips is equal to those obtained from the displacement field. It follows from Fig. (2) that the longitudinal ,lateral and shear total strain rates in the matrix strip are represented in terms of the continuum strain rate and fiber lateral and shear strain rates , as follows

$$\dot{e}_{11m} = \dot{e}_{11c}$$
$$\dot{e}_{22m} = \dot{e}_{22c}(1 + v) - \dot{e}_{22f}\,v \qquad\qquad (5)$$
$$\dot{e}_{12m} = \dot{e}_{12c}(1 + v) - \dot{e}_{12f}\,v$$

where the ratio of volume fractions, $v = v_f/v_m$. v_f is the volume fraction of the fibers and v_m is the volume fraction of the matrix. The matrix strains e_{11m}, e_{22m} and e_{12m} and the fiber strains e_{11f}, e_{22f} and e_{12f} are measured in directions parallel to the corresponding continuum strain components. The lateral and shear fiber strains have to satisfy the lateral and shear equilibrium on the fiber surfaces. For nonlinear constitutive fiber and matrix response, these stresses are obtained by integrating the equations of stress rates simultaneously with the constituents constitutive relationships. The stress tensor in both fibers and matrix is derived from their elastic strain tensor. Therefore, the stress rate tensor of fibers and matrix are derived from the corresponding elastic strain rate tensor which is the difference between the total and inelastic strain rate tensors. It can be shown that the components of the stress rate tensors of the matrix and the fibers are represented by the following differential equations

$$\dot{S}_{11m} = K_m[\dot{e}_{11c} - \dot{e}_{11m}^{i} + u_m(1+v)\dot{e}_{22c} - u_m\dot{e}_{22m}^{i} - vu_m\dot{e}_{22f}^{i}] - K_m v u_m \dot{e}_{22f}^{e}$$

$$= A1 + B1\ \dot{e}_{22f}^{e}$$

$$\dot{S}_{22m} = K_m\,[(1+v)\dot{e}_{22c} - \dot{e}_{22m} + u_m\,\dot{e}_{11c}^{i} - u_m\,\dot{e}_{11m}^{i} - v\,\dot{e}_{22f}^{i}]\ - K_m\,v\,\dot{e}_{22f}^{e}$$

$$= A2 + B2\ \dot{e}_{22f}^{e}$$

$$\dot{S}_{12m} = G_m\,[(1+v)\dot{e}_{12c}^{i} - \dot{e}_{12m}^{i} - v\,\dot{e}_{12f}^{i}]\ - G_m\,v\,\dot{e}_{12f}^{e}$$

$$= A3 + B3\ \dot{e}_{12f}^{e}$$

$$\dot{S}_{11f} = K_f\,[\,\dot{e}_{11c} - \dot{e}_{11f}^{i}\,] + K_f\,u_f\,\dot{e}_{22f}^{e}$$

$$= A4 + B4\ \dot{e}_{22f}^{e}$$

$$S_{22f} = S_{22m}$$

$$S_{12f} = S_{12m} \tag{6}$$

where $K_m = E_m/(1-u_m^{\,2})$ and $K_f = E_f/((1-u_f^{\,2})$ and E_m, G_m, u_m, E_f, G_f and u_f are the modulus, shear modulus and poisson ratio for the matrix and fibers respectively. The superscripts , **e** and **i**, refer to elastic and inelastic strain rates, respectively. The matrix stress components S_{11m}, S_{22m} and S_{12m} and the fibers stress components S_{11f}, S_{22f} and S_{12f} are parallel to the matrix and fiber strain components having the same indices. The rate of elastic lateral and shear strain in the fibers is obtained from the lateral and shear equilibrium between the matrix and fibers and can be written as follows

$$\dot{e}_{22f}^{e} = [A2/(K_f+vK_m)] - [u_f\,K_f/(K_f+vK_m)][\dot{e}_{11c} - \dot{e}_{11f}^{i}]$$
$$\dot{e}_{12f}^{e} = [A3\,/(G_f-B3)] \tag{7}$$

The inelastic strain rates and the rate of evolution of the resistance both in the matrix and fibers are obtained from Eqs. (1) and (2) upon using the appropriate material constants. The evaluation of local matrix and fiber stress state under strain control requires the simultaneous solution of eight differential equations ; the first four equations of the set of Eq. (6), Eq. (7), and two evolution equations of the constituents state variables. These equations reduce to five if the fibers are elastic.

Load Control

The internal in plane stress state of the composite lamina subjected to biaxial stress control shown in Fig. (1) can be obtained by assuming a continuum stress field for the composite. In this paper the field is assumed uniform and therefore the longitudinal , lateral and shear continuum stress rates under the biaxial stress rates shown in Fig. (2) are given by

$$\dot{S}_{11c} = \dot{S}_x \cos^2 \theta + \dot{S}_y \sin^2 \theta$$

$$\dot{S}_{22c} = \dot{S}_x \sin^2 \theta + \dot{S}_y \cos^2 \theta \qquad\qquad (8)$$

$$\dot{S}_{12c} = -(\dot{S}_x - \dot{S}_y) \sin \theta \cos \theta$$

where $\dot{S}_x$ and $\dot{S}_y$ are the applied stress rates in the **x** and **y** directions respectively. The continuum stresses S_{11c}, S_{22c} and S_{12c} are parallel to the continuum strains having the same indices. The lateral and shear stress in the fibers and matrix are equal to satisfy the equilibrium at fiber surfaces and are therefore assumed equal to the continuum lateral and shear stresses at any time. The longitudinal fiber stress is different from that of the matrix. The continuum longitudinal stress is an average stress that provides the same longitudinal traction in the composite. Therefore, the longitudinal stresses of fibers and matrix are related according to the rule of mixtures, $S_{11m} = (1+v) S_{11c} - v\, S_{11f}$. Notice that the longitudinal fiber and matrix total strain rates are equal. Therefore it follows that the fiber longitudinal stress rate and the stress state in fibers and matrix are given by

$$\dot{S}_{11f} = H1\, \dot{S}_{11c} + H2\, \dot{S}_{22c} + H3\, (\dot{e}^i_{11m} - \dot{e}^i_{11f})$$

$$S_{11m} = (1+v)\, S_{11c} - v\, S_{11f}$$

$$S_{22m} = S_{22f} = S_{22c}$$

$$S_{12m} = S_{12f} = S_{12c} \qquad\qquad (9)$$

where , $H1 = E_f [(1+v)/(E_m + vE_f)]$, $H2 = [(u_f E_m - u_m E_f)/(E_m + vE_f)]$ and $H3 = [E_m E_f/(E_m + vE_f)]$. Thus the evaluation of the stress state under stress control requires the simultaneous solution of three differential equations. The first one of the set of Eq. (9) and two evolution equations for the state variables of the fibers and the matrix. For elastic fibers, the three equations reduce to two.

EXPERIMENTAL PROGRAM

The material studied was a **4 ply**, unidirectional tungsten-fiber reinforced copper matrix composite. The volume fraction of the 0.008 inch diameter tungsten fiber was 9% and the orientation of the fibers was parallel to the load axis (i.e., θ=0). Load controlled fatigue experiments were performed at **260 C** in vacuum(<1.0E-5 Torr). Five specimens were tested to failure and two other tests were interrupted.

Fig. (3a) shows the typical cyclic behavior of the composite under cyclic tension-tension load control. The composite specimen ratchets continuously up to failure. Also, the range of cyclic strain decreases continuously with cycling and the hysteresis loops become nearly elastic at failure. Fig. (3b) is a plot of the maximum cyclic strain of six specimens as a function of the applied number of cycles. The ratcheting behavior is somewhat analogous to typical creep behavior of monolithic materials. The ratcheting rate is high at the beginning of the test and decreases to approach a steady state ratcheting rate after about **50-100** cycles. This steady state ratcheting regime spans over **1/4** of the life for the lowest load level and up to about **3/4** of the life for the highest load level. The ratcheting rate then increases as failure approaches. The failure strain is found to increase with increasing maximum cyclic stress ranging from **4.7 %** for a maximum stress of **35 ksi** to **12.7%** for a maximum stress of **41.1.ksi**.

Examination of the failed specimens revealed that fatigue cracks nucleated in the copper matrix via grain boundary cavitation. The cavities linked together to form cracks which grew around the fibers. However, some fibers broke before the composite specimens failed. No oxidation of the specimens was observed.

APPLICATION AND DISCUSSION

The material constants **C and D** should be evaluated from the fatigue data at the temperature of application. Such data were not available for tungsten and are not needed for the current application. The copper fatigue data at 260 C were not available either. However, room temperature high cycle fatigue data of copper was reported in [3]. The values of **C** and **D** at 260 C are assumed to be equal to those at room temperature, Table 1. The initial values of resistance **X0** for both copper and tungsten were obtained to provide a calculated transient tensile response close to the observed one. These values also are given in Table 1.

The fatigue shake down resistance of the insitu composite constituents may be different from the resistance of the individual constituents. Therefore, the shake down resistance of the matrix is evaluated from the composite fatigue data. Specimens **3,4** and **5** were used to develop a correlation between the matrix shake down resistance and applied stress. To develop such correlation, the tungsten fiber resistance is assumed to be constant and equal to its initial value. The matrix shake down resistance is obtained iteratively. Initially the matrix is assumed linear elastic and Eq. (9) is used to obtain an initial estimate for L_S of the matrix . This estimate is used in Eq. (3) to obtain an initial estimate for X_S of the matrix. Eq. (9) is then used to update L_S which is subsequently applied to Eq. (3) to update X_S. This process continues until the values of the L_S and X_S for the matrix converge. An effective stress is defined as the square root of the maximum value of **3*J2** for the local applied cyclic stress in the matrix. Upon correlating the shake down resistance of specimens **3,4** and **5** to their effective stress , it can be shown that the copper shake down resistance at **260 C** is almost constant and equal to **69.0 ksi**. This value for shake down is applied to predict the lives of the specimens **1,2,6,7**.

The calculated lives of these specimens are given in **Table 2**. Good agreement is evident between the observed and predicted low cycle fatigue specimens **3,4,5,6** and **7** . The agreement is not good for the high cycle fatigue specimens **1 and 2**. This can be attributed to the fact that the estimated value of $X_s=69.0$ is obtained from composite low cycle fatigue results which may not be appropriate for high cycle fatigue.

The variation of predicted life with fiber orientation is shown in Fig. (4) for the same stresses as had been applied to specimen 7. This prediction is based on the assumption that composite fatigue failure remains matrix-controlled and that the matrix fails due to inelasticity micromechanisms. Notice that the effective stress and L_s in the matrix decreases as the angle of fiber orientation increases up to an angle equal approximately to 33 degrees after which they monotonically increase as shown in Fig. (4). The decrease in L_s is followed by a predicted increase in matrix life. The composite life at any given angle of fiber orientation , however, could be fiber-controlled or fiber/matrix interface-controlled. An appropriate life prediction requires the predicted lives of each potential contributor on Fig. (4). The composite life at any orientation would be the lowest of the predicted constituent lives.

FUTURE RESEARCH

Three recommendations for future research are proposed. First, the analysis should be developed to account for the effect of inelasticity due to thermal strain. Such a step is mandatory if the model is to be applied to predict thermomechanical fatigue life. Second, it is required to conduct higher cyclic life composite fatigue experiments to improve the representation of the shake down resistance for high cycle fatigue. Third, composite fatigue experiments should be conducted at different angles of orientation to verify the validity of the proposed life prediction method.

REFERENCES

1) Johnson, W.S., "Fatigue Testing and Damage Development in Continuous Fiber Reinforced Metal Matrix Composites," NASA T.M. 100628, June, 1988.
2) Abuelfoutouh, N.M., "Prediction of Isothermal Low Cycle Fatigue Response And Associated Life of High Temperature Isotropic Materials, " submitted for publication in the 'Symposium of Plasticity 89', Tsu, Japan, July 1989.
3) Lukas, P. and Kunz, L., "Effect of Grain Size on The High Cycle Fatigue Behavior of Polycrystalline Copper," Mat. Science and Engrg., vol. 85, 1987, pp. 67-75.

TABLE 1 - MATERIAL CONST				
	units		Cu	W
A0	1/sec Ksi		10000	10000
r			7.0	15.0
C	1/Ksi		-.2526	--
D			-37.2	--
x0	Ksi		23.0	800.

TABLE 2 - FATIGUE RESULTS (0=0)				
No	Max strs Ksi	Min strs Ksi	N_f exp	N_f theo
1	33.5	1.42	30419*	28430
2	34.3	1.53	62403*	18510
3	35.0	1.48	12588	12370
4	37.2	1.53	3117	4648
5	38.6	1.52	2727	2464
6	40.2	1.49	1864	1250
7	41.1	1.51	933	855

* Test interrupted prior to failure

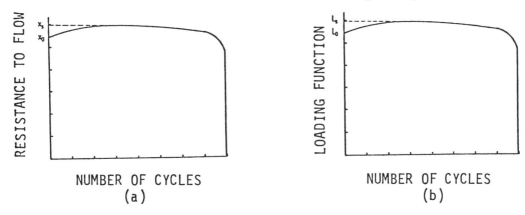

Figure-1- Schematic variation of the resistance of material to flow and of loading function with the number of applied fatigue cycles.

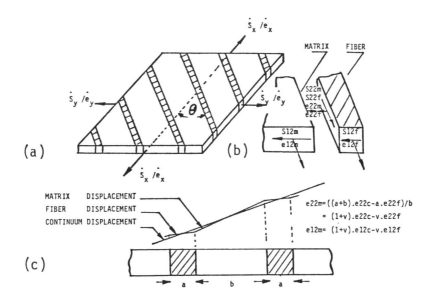

Figure-2- The composite lamina, (a) Applied biaxial stress and strain rates. (b)-Fiber and matrix in plane stresses and strains. (c)-Fiber, matrix and continuum strains and displacements in the direction perpendicular to fibers.

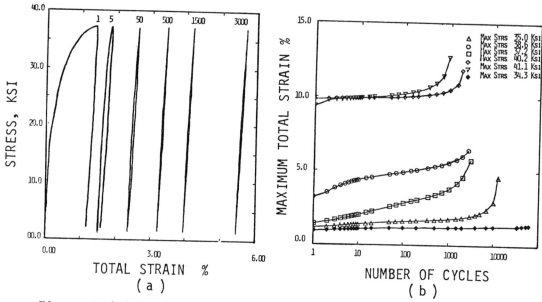

Figure-3-(a) Typical stress-strain behavior of 9% W reinforced Cu
under load controlled fatigue at 260 C. Curve lables indicate
cycle numbers. (b) Maximum cyclic strain versus number of cycles
for the same composite at six levels of load controlled fatigue.

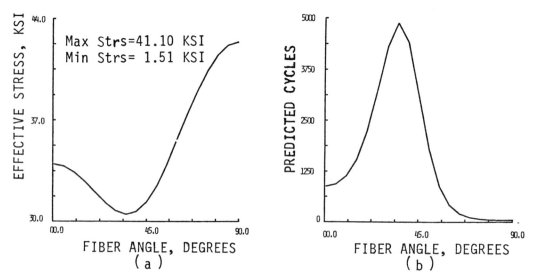

Figure-4- Variation of matrix effective stress with fiber angle
(b) Variation of matrix predicted life with fiber angle.

130

Young's Modulus Measurement of Textron's Silicon Carbide and Boron Fibers

DAVID GIANNELLI

Abstract

The measurement of Young's Modulus on Textron Specialty Materials' Silicon Carbide and Boron fibers required the development and evaluation of a technique which would assure accurate and repeatable strain readings. Three test techniques were used to generate data, and these data were compared to theoretical values determined analytically. One set of test data was acquired from load vs. time plots and corrected for system compliance, a second set of data was acquired using a laser based strain sensor to measure strain directly from targets bonded to the test fiber, and a third set of data was generated using a Sonic Modulus test method which measures the velocity of sonic pulses through the test fiber.

This paper describes in some detail the test techniques and analytical techniques used and includes the data generated in these experiments.

Test Methods

Load Versus Time Method

This method involved the testing of multiple test specimens at each of seven gage lengths as defined by grip separation. By driving the moving crosshead of an Instron Universal Tester at a constant velocity and synchronously driving the chart at a fixed rate, we produced a series of behavior profiles with load on one coordinate and apparent deflection or strain on the other. By using different gage lengths, we can determine the real compliance of the test system. This allows the calculation of a compliance factor that can then be applied to all the data generated to correct for the differences in gage length in calculating Young's Modulus.

D. Giannelli, Test Engineer, Textron Specialty Materials
2 Industrial Avenue, Lowell, MA 01851

This experiment consisted of testing five specimens at each of seven gage lengths; one inch, two inch, three inch, five inch, seven inch, ten inch and twenty inches. All test specimens were taken from the same spool of SCS-9 3.2 mil diameter silicon carbide fiber. An Instron Model 1123 Universal Testing Machine was used for all tests and incorporated a 200 pound capacity load cell and pneumatic grips with steel faces. Grip air pressure was maintained at 45 psi, crosshead velocity was 0.05 inches per minute and chart speeds of two inches and five inches per minute were used. Aluminum foil (.003 inch thick) was used to line each grip face to enhance grip quality and minimize grip induced failure of the test specimens. Uncorrected modulus data is shown in Table I.

The following calculation is used to determine Young's Modulus for the Load Vs. Time Method.

$$Y_m = \sigma / \epsilon \tag{1}$$

Where:

Y_m = Young's Modulus (psi)
σ = Stress at a given load (psi)
ϵ = Strain at a given load (in/in)

The system compliance factor was calculated using the technique described in ASTM D3379 titled "Tensile Strength and Young's Modulus for High Modulus Single Filament Materials". The compliance factor established was applied to the individual data points producing the adjusted Young's Modulus data shown in Table II.

The indicated compliance factor was calculated using the following equation:

$$C_a = \frac{I}{P} \times \frac{H}{S} \tag{2}$$

where I = total extension of the linear section of the load time curve extrapolated to full scale load (inches)
    H = cross head velocity (inches/minute)
    P = full scale load (pounds)
    S = chart speed (inches/minute)

Results were plotted on an Indicated Compliance versus Gage Length graph and a best fit line was drawn through the data points to the zero gage length intercept. The vertical off set from zero gage length is the system compliance (Cs) in inches per pound. Data is plotted in Figure I.

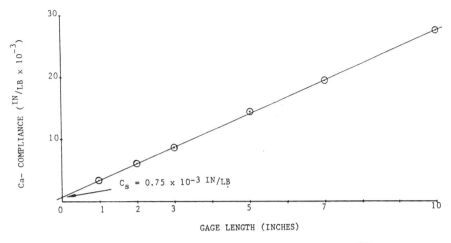

FIGURE 1. COMPLIANCE FOR MODEL 1123 INSTRON PER ASTM D-3379

The true compliance is then determined by subtracting the system compliance from the indicated compliance.

$$C = C_a - C_s \qquad (3)$$

Where: C = True Compliance (in/lb)
C_a = Indicated Compliance (in/lb)
C_s = System Compliance (in/lb)

Adjusted Young's Modulus can be calculated using:

$$Y_m = L/CA \qquad (4)$$

Where: Y_m = Young's Modulus (psi)
L = Specimen gage length (in)
C = True compliance (in/lb)
A = Average cross-sectional area (in^2)

Direct Strain Measurement Method (Laser Based Strain Sensor)

The accurate measurement of Young's Modulus of SCS-9 (3.2 mil) Silicon Carbide Fiber required the development and evaluation of a targeting technique that would assure accurate repeatable strain readings using a laser based strain sensing system.

Initial experiments involved the testing of eight specimens using the same fixturing that was used in the gage length study and a grip separation of five inches. Aluminum tabs were bonded to the fiber to establish a 0.250 inch gage length over which any axial deformation would be sensed by the laser system. During these tests, a load versus strain record was acquired by the laser system and a simultaneous load versus time record was plotted on the Instron recorder.

The data acquired from the Instron recorder indicated a Young's Modulus value of 46.9 msi with a coefficient of variation of 2% using equation (4) and the laser system data indicated an average modulus of 47.3 msi and a coefficient of variation of 41% using equation (1). The wide scatter of the data acquired from the laser system caused a more detailed examination of possible problem areas. This evaluation revealed three potential problem areas; a floating upper grip, uneven edges on the reference surface of the targets and the laser system not in close enough proximity to the central axis of the test specimen. Each of these problem areas were addressed and a second set of experimental data was generated. The upper grip was fixed in place, the test specimen and laser system were precisely centered and the targets were made from steel shim stock to provide better edges to sight on.

Eighteen additional tests were performed with the data being acquired in the same way as in the initial experiment. Four different gage lengths were used to validate the laser system with five tests at 0.25 and 1.0 inch gage lengths and four tests at 0.50 and 0.75 inch gage lengths. The data acquired from the load-time curves indicated a Young's Modulus of 47.1 msi with a coefficient of variation of 2% and the laser system produced an average Young's Modulus of 48.4 msi with a coefficient of variation of 9%. A statistical comparison of the mean values indicates minimal statistical difference between them. These data are presented in Tables III and IV.

Sonic Modulus

Sonic Modulus was chosen as a third method for measuring the elastic modulus of SCS-9 fiber as a means of verifying the values attained in the mechanical modulus experiments. Three separate samples of 3.2 mil SCS-9 fiber, 20 feet long were tested. This method consisted of measuring the time in seconds a sonic pulse took to transmit through a fiber over a given distance. The Sonic Young's Modulus calculation is:

$$Y_m = P_f \, C^2 * 1.45 \times 10^5 \qquad (5)$$

Where:
$$Y_m = \text{Young's Modulus (psi)}$$
$$P_f = \text{density of fiber (gms/cc)}$$
$$C = \text{Sonic Velocity (km/sec)}$$
$$1.45 \times 10^5 = \text{Conversion factor from dynes/cm}^2 \text{ to psi}$$

The series of sonic modulus tests were conducted using the same value for density as is used in determining the sonic modulus of our 5.6 mil SiC fiber since the density of the 3.2

mil fiber was unknown. The uncorrected modulus values averaged 52 msi which is midway between values normally seen on the 5.6 mil fiber and modulus values measured mechanically on the SCS-9 fiber in the two previous experiments. A theoretical density had to be calculated for the 3.2 mil fiber. We can use rule of mixtures calculations based on the density of the SiC sheath to determine the density of the 3.2 mil SCS-9 fiber in the following manner.

Density of the SiC sheath is calculated as follows:

$$P_s = \frac{P_f - V_c P_c}{V_s} \tag{6}$$

Where: P_s = density of SiC sheath (gms/cc)
P_f = density of 5.6 mil fiber = 3.05 gms/cc
V_c = volume represented by carbon core = 6%
P_c = density of carbon core = 1.8 gms/cc
V_s = volume represented by the SiC sheath = 94%

Therefore: P_s = 3.13 gms/cc

Density of 3.2 mil SCS-9 fiber can be calculated as follows:

$$P_f = V_s P_s + V_c P_c \tag{7}$$

Where: P_f = density of 3.2 mil SCS-9 fiber (gms/cc)
V_s = volume represented by SiC sheath = 81%
P_s = density of SiC sheath = 3.13 (gms/cc)
V_c = volume represented by carbon core = 19%
P_c = density of the carbon core = 1.8 (gms/cc)

Therefore: P_f = 2.88 gms/cc

The theoretical density of the SCS-9 fiber was applied to the "as acquired" sonic modulus data and produced an average modulus of 48.4 msi. This value correlates very well with the mechanical modulus values of 48.0 and 48.4 msi using the two previously described methods.

Theoretical Modulus Calculations

Rule of Mixtures Calculation

A maximum theoretical modulus of a composite material can be determined by applying the rule of mixtures. If we use the same percentages that were used in the cross-section calculations and use a modulus of 6 msi for the carbon core and 60 msi for the SiC sheath, we can calculate the modulus of the SCS-9 fiber as follows:

$$Y_m = Y_c V_c + Y_s V_s \qquad (8)$$

Where: Y_m = Young's Modulus of 3.2 mil fiber (msi)
Y_c = Young's Modulus of carbon core = 6 msi
V_c = Volume represented by the carbon core = 19%
Y_s = Young's Modulus of SiC sheath = 60 msi
V_s = Volume represented by the SiC sheath = 81%

Therefore: $Y_m = 49.7$ msi

Theoretical modulus calculation based on fiber cross-section.

This method of estimating the Young's Modulus of 3.2 mil SCS-9 fiber is based on the rational that the ratio of the SiC sheath of the two fibers is the major contributing constituent.

The equation derived from this ratio is as follows:

$$Y_{3.2} = \left\{ 1 - (A_{5.6} - A_{3.2}) \right\} * Y_{5.6} \qquad (9)$$

Where: $Y_{3.2}$ = Young's Modulus of 3.2 mil SCS-9 fiber, (psi)
$A_{5.6}$ = Area represented by the SiC sheath of the 5.6 mil SCS fiber = 93.7%
$A_{3.2}$ = Area represented by the SiC sheath of the 3.2 mil SCS fiber = 81%
$Y_{5.6}$ = Young's Modulus of 5.6 mil SCS fiber = 56 msi

Therefore: $Y_{3.2} = 48.3$ msi

Conclusions

We have approached the definition of the Young's Modulus of SCS-9 3.2 mil silicon carbide fiber from five different directions and seem to find that each method tends to reinforce the validity of the data acquired. A simplified presentation of the modulus data indicates that the mean values of the modulus obviously are from a very tightly distributed population.

A summary of the data using the previously described methods is presented as follows:

Method	Average Modulus (msi)
Load versus Time Method	48.0
Load versus Strain Method	48.4
Sonic Method	48.4
Cross Section Method	48.9
Rule of Mixtures Calculation	49.7
Overall Statistics:	x = 48.6
	s = 0.61
	cv = 1.2%

The extremely small scatter of these mean values indicates that the modulus of SCS-9 3.2 mil fiber is 48.5 ±1 msi.

Table I

Uncorrected Young's Modulus of SCS-9 Silicon Carbide Fiber
Load Versus Time Method
Modulus ($x10^6$ psi)

Gage Length	Run 1	Run 2	Run 3	Run 4	Run 5	Avg.
1 inch	40.9	32.4	34.5	39.8	39.6	37.4
2 inch	41.2	42.6	43.6	44.2	43.8	43.1
3 inch	43.2	44.2	46.1	42.7	41.3	43.5
5 inch	45.8	46.2	45.3	46.1	45.6	45.8
7 inch	44.6	46.0	47.0	45.8	45.6	45.8
10 inch	47.5	46.0	46.7	46.0	46.2	46.5
20 inch	46.7	46.3	46.0	-	-	46.3

Overall Statistics:
x = 43.9
s = 3.45
cv = 8%
R = 32.4 - 47.5
N = 33

Table II

Young's Modulus Values of SCS-9 Silicon Carbide Fiber
Load Versus Time Method with Compliance Factor Applied

Modulus ($x10^6$ psi)

Gage Length	Run 1	Run 2	Run 3	Run 4	Run 5	Avg.
1 inch	53.7	40.0	43.2	52	51.5	48.1
2 inch	46.9	48.7	50.0	50.8	50.2	49.3
3 inch	47.1	48.3	50.6	46.6	44.9	47.5
5 inch	48.3	48.8	47.8	48.7	48.2	48.4
7 inch	46.4	47.9	49.0	47.7	47.6	47.7
10 inch	48.9	47.3	48.0	47.3	47.5	47.8
20 inch	47.3	47.0	46.7	-	-	47.0

Overall Statistics:
x = 48.0
s = 2.47
cv = 5.1%
Range = 40.0 - 53.7
N = 33

Table III

Young's Modulus of SCS-9 Fiber Using Laser Strain Sensor
for Direct Measurement

Modulus (x10^6 psi)

Specimen #	Load Vs. Time Uncorrected	Load Vs. Time Corrected	Zygo Laser Sensor
1	44.0	46.5	40.8
2	45.9	48.6	31.2
3	45.6	48.3	93.4
4	43.8	46.3	35.5
5	43.7	46.1	49.8
6	44.5	47.0	48.9
7	43.6	46.0	40.5
8	43.8	46.3	38.5

Overall Statistics:

x =	44.4	46.9	47.3
s =	.90	1.01	19.6
cv =	2%	2%	41%
range =	43.6-45.9	46.0-48.6	31.2-93.4
N =	8	8	8

Table IV

Young's Modulus of SCS-9 Fiber Using Laser
Strain Sensor for Direct Measurements

Modulus (x10^6 psi)

Gage Length	Run 1	Run 2	Run 3	Run 4	Run 5	Avg.	CV
0.25	56.4	48.4	47.9	36.2	45.2	46.8	15.5%
0.25	(47.2)	(47.2)	(47.6)	(47.6)	(46.7)	(47.3)	(0.8%)
0.50	46.3	48.1	47.0	45.3	-	46.7	2.5
0.50	(48.1)	(46.3)	(46.0)	(44.7)	-	(46.3)	(3.0)
0.75	45.3	48.8	50.2	55.9	-	50.0	8.8%
0.75	(46.9)	(47.2)	(46.2)	(45.7)	(45.7)	(46.6)	(1.4%)
1.00	48.4	50.0	49.9	50.7	50.5	49.9	1.8%
1.00	47.5	(48.3)	(47.6)	(48.9)	(48.0)	(48.1)	(1.2%)

Data in parentheses ( ) corrected load vs. time data.

Overall Statistics:

	Laser Sensor	Load Vs. Time
x =	48.4	47.1
s =	4.36	1.0
cv =	9%	2%
Range =	36.2-56.4	44.7-48.9
N =	18	19

Thermal Stresses in Coated Fiber Composites

TUNGYANG CHEN[1] **AND**
GEORGE J. DVORAK[2]

YAKOV BENVENISTE[3]

ABSTRACT

This paper presents a micromechanical analysis of stress fields in coated and uncoated fiber composites subjected to both mechanical and thermal load fluctuations. The analysis of unidirectional materials is based on the 'average stress in the matrix' concept of Mori and Tanaka (1973) [6]. In the present work, the concept is extended to coated fibers. The advantage of this approach is that it permits one to introduce exact elasticity solutions of the coated fiber problem into the analysis. Such relations were derived for the case of distinct transversely isotropic phases. All possible mechnaical loadings, and a uniform temperature change were considered. Results are given for several specific systems.

INTRODUCTION

Micromechanics analyses of composite materials are often related to Eshelby's (1957) [4] result that the strain field in an ellipsoidal inclusion bonded to a uniformly strained infinite medium is also uniform. This result is commonly used to evaluate overall properties and average local fields in composite aggregates in terms of the phase stress and strain concentration tensors. Unfortunately, local fields in coated inclusions are generally not uniform, hence the phase concentration tensors cannot be easily evaluated. Therefore, the analysis of composite reinforced by coated fibers or particles is one of the more difficult problems in micromechanics.

Walpole (1978) [8] was among the first to consider problems of this kind. He assumed that a very thin coating has no effect on strain distribution in the inclusion. Hatta and Taya (1986) [5] considered the heat conduction problem in composites reinforced by short coated fibers in the context of the original Mori-Tanaka method. Recently, Pagano and Tandon (1988) [7] analyzed a multidirectional coated fiber composite by means of a three-phase concentric cylinder model.

The present paper is concerned with evaluation of local fields and overall properties of composites reinforced by coated fibers . The results are derived from a variant of Benveniste's (1987) [1] reexamination of the Mori-Tanaka's method. In particular, the local fields in a coated inclusions are approximated by those found when the coated inclusion is embedded in an unbounded matrix medium subjected to the average matrix stresses at infinity. The advantage of this approach is that the local fields in the coating and inclusion,

---

[1] Graduate Research Assistant, Dept. of Civil Eng., R.P.I.,Troy, N. Y.
[2] Professor, Dept. of Civil Eng., R.P.I.,Troy, N. Y.
[3] Professor, Tel-Aviv University, ISRAEL

and in the matrix can be evaluated by using the solution for a single coated fiber in an infinite matrix. Specific results are found for systems reinforced by aligned coated fibers, where the phases are isotropic or transversely isotropic elastic solids. We also consider the possibility of plastic yielding in the unidirectional composites.

STRESS AND STRAIN FIELDS IN THREE-PHASE COMPOSITES

Consider a three-phase composite material consisting of a continuous matrix phase m, in which there are embedded a fiber phase f, and a third phase g which represents a layer of coating that encapsulates each fiber of the f phase. The fibers have a circular crosssection. The phases are assumed elastic and perfectly bonded during deformation. The composite medium is statistically homogeneous. The thermoelastic constitutive equations of the phases are given in the form

$$\sigma_r = \mathbf{L}_r \varepsilon_r + \mathbf{l}_r \theta \tag{1}$$

$$\varepsilon_r = \mathbf{M}_r \sigma_r + \mathbf{m}_r \theta \tag{2}$$

where $r = f, g, m$; $\mathbf{L}_r$ and $\mathbf{M}_r = (\mathbf{L}_r)^{-1}$ are the stiffness and compliance tensors; $\mathbf{l}_r$ is the thermal stress vector and $\mathbf{m}_r$ is the thermal strain vector of the expansion coefficients, such that

$$\mathbf{l}_r = -\mathbf{L}_r \mathbf{m}_r \tag{3}$$

Define the following thermomechanical loading problems

$$\sigma_n(S) = \sigma_o \mathbf{n} \qquad\qquad \theta(S) = \theta_o \tag{4}$$

where $\sigma_n(S)$ is the traction at the external boundary S of a representative volume V of the composite under consideration, $\mathbf{n}$ denotes the exterior normal to the surface S; σ_o is the applied uniform stress field; $\theta(S)$ is the temperature rise at S, and θ_o is a constant quantity.

The composite is subjected to boundary conditions (4). The solution for the stress field in the phases can be expressed in the form:

$$\sigma_r(\mathbf{x}) = \mathbf{B}_r(\mathbf{x}) \sigma_o + \mathbf{b}_r(\mathbf{x}) \theta_o \qquad\qquad r = f, g, m \tag{5}$$

where $\mathbf{B}_r(\mathbf{x})$ and $\mathbf{b}_r(\mathbf{x})$ are fourth and second order tensors, respectively. Their volume averages $\mathbf{B}_r$ and $\mathbf{b}_r$ are usually referred to as mechanical and thermal stress concentration factors.

In the Mori-Tanaka method,, the stress fields in phases f and g are assumed to be equal to the fields in a single coated fiber which is embedded in an unbounded matrix medium and subjected to remotely applied stresses which are equal to the yet unknown average stress in the matrix. Also a uniform temperature change is applied.

Suppose that the single inclusion is surrounded by a large matrix volume V' with surface S', Fig 1. The solutions of this dilute problem assumes the form

$$\sigma_r(x) = W_r(x) \sigma_m + w_r(x) \theta_o \qquad\qquad r = f, g \qquad (6)$$

we will solve this auxiliary problem in the next section. To determine σ_m, we recall that the overall uniform stress and the local average stress are connected by the relations

$$\sum_r c_r \sigma_r = \sigma_o \qquad\qquad r = f, g, m \qquad (7)$$

where c_r denote the phase volume fractions, $c_f + c_g + c_m = 1$. From equations (6) into (7), one finds that the unknown matrix stress is equal to

$$\sigma_m = \left[\sum_r c_r W_r \right]^{-1} \left[\sigma_o - \theta_o \sum_r c_r w_r \right] \qquad (8)$$

This opens the way for evaluation of the partial stress concentration factors in (6) and of the average matrix stress (8). One can then obtain the solution (5).   An entirely similar procedure can be applied under strain boundary conditions.

The effective thermoelastic constitutive relations of the composite medium are defined as:

$$\sigma = L \varepsilon + l \theta \qquad (9)$$

$$\varepsilon = M \sigma + m \theta \qquad (10)$$

where σ, ε and θ denote representative volume averages and L, M, l, m are overall stiffness, compliance, thermal stress and thermal strain tensors, respectively. Using equations (6), (7), and (8), one can derive :

$$M = \left[\sum_r c_r M_r W_r \right] \left[\sum_r c_r W_r \right]^{-1} \qquad (11)$$

$$m = \left[\sum_r c_r M_r W_r \right] \left[\sum_r c_r W_r \right]^{-1} \left[-\sum_r c_r w_r \right] +$$

$$\sum_r c_r (M_r w_r + m_r) \qquad (12)$$

We note here that Benveniste et al (1989) [2] has proved that the results are consistent in that the overall compliance tensor is the inverse of stiffness tensor. Also, we have numerically verified that the effective stiffness L and compliance M are sysmetric.

AUXILIARY PROBLEMS

In the evaluation of $\mathbf{W}_r$ and $\mathbf{w}_r$, the composite is subjected to several traction boundary conditions and to a uniform temperature change. We wish to find the stress distribution in the fiber, coating and matrix which surrounds the periphery of the coated fiber. The loading cases are shown in Fig. 2 :

Case 1 - Transverse hydrostatic stress
Case 2 - Transverse shear stress
Case 3 - Transverse normal stress
Case 4 - Axial normal stress
Case 5 - Longitudinal shear stress
Case 6 - Uniform change in temperature

Case 3 can be obtained as a superposition of cases 1 and 2. In what follows, the cylindrical coordinates r, θ, z are used; the respective stress or strain components are always written as subscripts, while the phase designation is always written as a supercript.

(i) Cases 1,4,6

The solution of the above boundary value problems can be derived from the following admissible displacement field:

$$u_r^f = A_f r \qquad\qquad u_r^m = A_m r + B_m / r$$
$$u_r^g = A_g r + B_g / r \qquad u_z^r = \varepsilon_z^o \qquad\qquad (13)$$

where $u_r^{(i)}$, with the superscript (i)=f,g,m, are the radial displacements in the phases; $u_z^{(i)}$ denotes the axial displacement in the z direction . The constants A_f, A_g, A_m, B_g, B_m and ε_z^o are to be determined from the conditions of continuity of displacements and tractions at the two interfaces, and from the traction boundary conditions at infinity.

(ii) Case 2

The admissible displacement field of this problem has the form given by Christensen and Lo (1979) [3] :

$$u_r^f = (b\,\sigma^o / 4\,\mu_f) \left[a_1 (\eta_f - 3) \left[\frac{r}{b} \right]^3 + d_1 \left[\frac{r}{b} \right] \right] \cos 2\theta \quad , \quad 0 \le r \le a \qquad (14)$$

$$u_\theta^f = (b\,\sigma^o / 4\,\mu_f) \left[a_1 (\eta_f + 3) \left[\frac{r}{b} \right]^3 - d_1 \left[\frac{r}{b} \right] \right] \sin 2\theta \quad , \quad 0 \le r \le a \qquad (15)$$

$$u_r^g = (b\,\sigma^o / 4\,\mu_g) \left[a_2 (\eta_g - 3) \left[\frac{r}{b} \right]^3 + d_2 \left[\frac{r}{b} \right] + c_2 (\eta_g + 1) \left[\frac{b}{r} \right] + b_2 \left[\frac{b}{r} \right]^3 \right] \cos 2\theta \qquad (16)$$

$$u_\theta^g = (b\,\sigma^o / 4\mu_g) \left[a_2(\eta_g+3)\left[\frac{r}{b}\right]^3 - d_2\left[\frac{r}{b}\right] - c_2(\eta_g-1)\left[\frac{b}{r}\right] + b_2\left[\frac{b}{r}\right]^3 \right] \sin 2\theta \qquad (17)$$

$$u_r^m = (b\,\sigma^o / 4\mu_m) \left[2\left[\frac{r}{b}\right] + (\eta_m+1)\,a_3\left[\frac{b}{r}\right] + c_3\left[\frac{b}{r}\right]^3 \right] \cos 2\theta \qquad (18)$$

$$u_\theta^m = -(b\,\sigma^o / 4\mu_m) \left[2\left[\frac{r}{b}\right] + (\eta_m-1)\,a_3\left[\frac{b}{r}\right] - c_3\left[\frac{b}{r}\right]^3 \right] \sin 2\theta \qquad (19)$$

$$u_z^{(i)} = 0 \qquad (20)$$

where a and b denote the inner and outer radii of the coating, μ_r are phase shear moduli, and $\eta_r = 3 - 4\,v_r$; v_r denotes the Poisson's ratio; a_i, b_i, c_i, and d_i are unknown constants to be determined from the interface conditions.

(iii) Case 5

The general displacement field of anti-plane shear is

$$u_z^f = A_f\,r\,\sin\theta \qquad\qquad u_z^g = \left(A_g\,r + \frac{B_g}{r} \right)\sin\theta$$

$$u_z^m = \left(A_m\,r + \frac{B_m}{r} \right)\sin\theta \qquad u_x^r = u_y^r = 0 \qquad (21)$$

The five constants A_f, A_g, A_m, B_g, B_m are obtained from continuity of the u_z displacement and the σ_{rz} stress at the interfaces, as well as from the boundary condition $\sigma_{yz} = \sigma^o$ at $r = \infty$.

NUMERICAL RESULTS

The stress distributions in coated and uncoated composites are illustrated for several systems. The fiber volume fraction is assumed to be 0.4 .

Table 1 shows the thermal stresses in the fiber, coating and matrix under uniform temperature change of 1°C. Since the volume fraction of the coating is very small (less than 1%), the results for coated fiber composites are not much different from those for the uncoated fiber composites. However, very different stress magnitudes are found in different systems.

Figure 3 illustrates the thermal stress distribution in the radial direction for the system consisting of SiC fiber, carbon coating and Ti$_3$Al matrix. The coating thickness is 1 μm; fiber radius is 75 μm. The fiber stresses are uniform in this case. Table 2 presents the average thermal stress caused by cooling from the processing temperature to the room temperature, for 4 different uncoated systems. Figures 4 and 5 show yield stress and the effective stress vs temperature during cooling in a system consisting of an Al_2O_3 fiber, in a Ti_3Al or Ni_3Al matrix. These results indicate that yielding may take place during cooling.

ACKNOWLEDGEMENTS

Support for this work was provided by the DARPA - HiTASC program at R.P.I. .

REFERENCES

1. Benveniste, Y.,1987, "A New Approach to The Application of Mori-Tanaka's Theory in Composite Materias, " Mechanics of Materials, Vol. 6, pp. 147-157.

2. Benveniste, Y., Dvorak, G. J. and Chen, T., 1989, "Stress fields in Composites with coated Inclusions, " Mechanics of Materials, in press.

3. Christensen, R. M. and Lo , K. H.,1979, " Solutions for the Effective Shear Properties of Three-Phase Sphere and Cylinder Models, " Journal of Mechanics and Physics of Solids, Vol. 27, pp. 315.

4. Eshelby, J. D.,1957, " The Determination of the Elastic Field of an Ellipsoidal Inclusion and Related Problems, " Proceedings of the Royal Society, London, series A, Vol. 241, pp. 376-396.

5. Hatta, H and Taya, M.,1986, "Thermal Conductivity of Coated Filler Composites, "Journal of Applied Physics, Vol. 59, pp. 1851-1860.

6. Mori, T. and Tanaka K.,1973, " Average Stress in Matrix and Average Elastic Energy of Materials With Misfitting Inclusions, " Acta Metallurgica, Vol. 21, pp. 571-574.

7. Pagano, N. J. and Tandon, G. P.,1988,"Elastic Response of Multi-directional Coated-fiber Composites," Composite Science and Technology, Vol.31, pp.273-293.

8. Walpole, L. J., 1978, "A Coated Inclusion in an Elastic Medium," Mathematical Proceeding Camb. Phil. Society, Vol. 83, pp. 495-506.

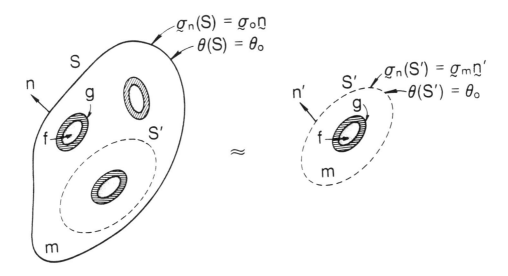

Fig. 1   A schematic representation of Mori–Tanaka's method for thermoelastic problems

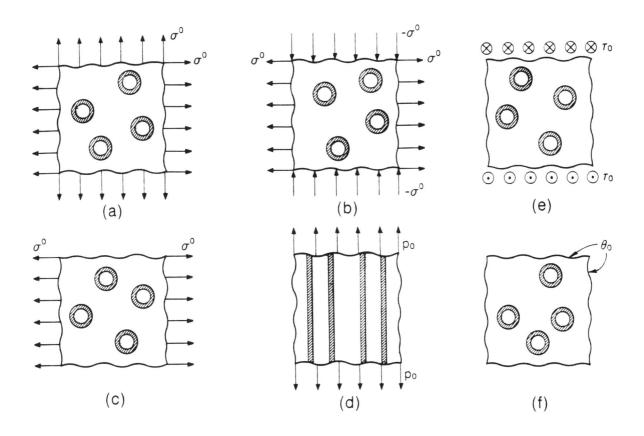

Fig. 2   Mechanical and thermal loading configurations

MPa/°C	$\bar{\sigma}_{rr}^{c}$	$\bar{\sigma}_{\theta\theta}^{c}$	$\sigma_{\theta\theta}^{f}$	$\sigma_{\theta\theta}^{m}$
Nicalon f. Carbon c. LAS matrix	−0.0405 (−0.03)	−0.2205	0.0218 (0.0246)	−0.043 (−0.058)
Graphite f. Yttria c. SiC matrix	2.673 (3.069)	2.577	2.923 (3.074)	−6.855 (−7.079)
Tungsten f. Carbon c. Nickel matrix	0.769 (0.7981)	0.288	0.771 (0.797)	−1.814 (1.848)
SiC f. Carbon c. Ti$_3$Al m.	0.1936 (0.1980)	0.1064	0.1942 (0.1980)	−0.4516 (−0.4523)

() ⇒ without coating

Table 1   Average thermal stresses of uniform temperature change of 1°C

MPa	PT(°C) YT(°C) ΔT_e(°C)	STRESS EVALUATION ΔT(°C)	$\bar{\sigma}_{rr}^{f}$	$\bar{\sigma}_{rr}^{m}$	$\bar{\sigma}_{\theta\theta}^{f}$ interface	$\bar{\sigma}_{\theta\theta}^{m}$ interface	σ_{xx}^{f}	σ_{xx}^{m}
AL$_2$O$_3$ Ti$_3$Al	950 X −930	−100	−12.91	−6.01	−12.91	26.82	−40.07	21.58
SCS − 6 Ti$_3$Al	950 733.4 −216.6	−100	−43.19	−26.69	−43.19	89.72	−139.43	75.08
SCS − 6 Ni$_3$Al	1200 1098 −102	−100	−73.85	−34.35	−73.85	153.39	−207.19	111.55
Al$_2$O$_3$ Ni$_3$Al	1200 977.7 −222.3	−100	−46.75	−21.75	−46.75	97.11	−129.61	67.79

Table 2   Average thermal stresses after cooling 100°C

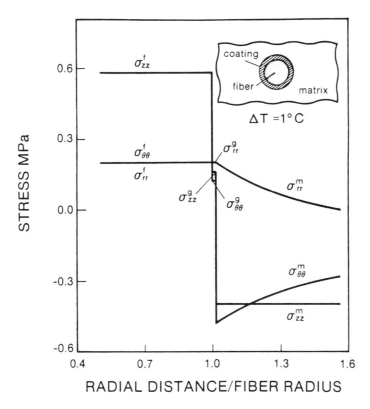

Fig. 3   Stress distribution for uniform temperature change of 1°C

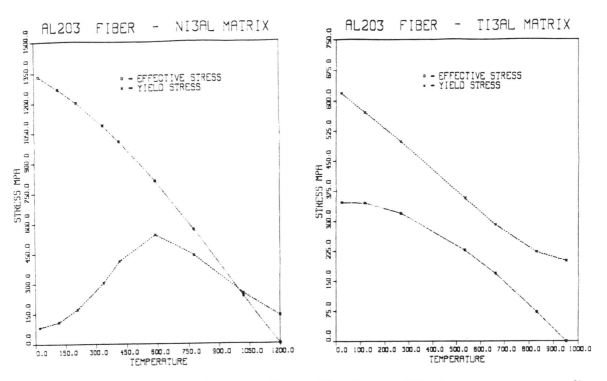

Fig. 4   Effective stress during cooling   Fig. 5   Effective stress during cooling

Effect of Interfacial Properties on Matrix Cracking Stress of Fiber Reinforced Ceramics

R. A. SHIMANSKY AND H. T. HAHN

ABSTRACT

Subcritical damage in continuous fiber brittle matrix composites is governed by fiber pullout mechanisms that depend on characteristic strength of the fiber/matrix interface. A predictive tool that approximates the effect of interface properties on matrix cracking stress has been developed as a guide for the development and optimization of interface modified composites. Based upon an existing fiber pullout model, an approximate expression for matrix cracking stress is derived. The effects of thermal mismatch, interfacial friction and interfacial debonding energy are studied for two brittle matrix composites: 35% Nicalon®/CAS and 30% SCS-6/RBSN. The predicted results, based upon assumed interfacial properties, predict matrix cracking stress values that agree with experimentally measured values.

INTRODUCTION

Reinforcement of ceramics and other brittle materials with continuous fibers has resulted in composites with significantly higher toughness than similar monolithic matrix materials. Fibers weakly bonded to the matrix increase composite toughness by permitting sub-critical matrix fracture, interface debonding and fiber sliding. These microscopic damage mechanisms result in large macroscopic strains prior to ultimate failure making the composite fail in a graceful manner. From a macroscopic standpoint, the mechanical performance of these composites made of inherently brittle constituents resembles that of clastic-plastic strain hardening metals. Improved toughness has been observed in both reinforced concrete and more recently in advanced composites such as glasses and glass-ceramics reinforced with carbon and SiC fibers[1-5].

To understand the fracture processes that occur in these materials, consider a unidirectional fiber composite loaded in simple tension parallel to the fibers. Subcritical failure initiates with the development of one or more matrix cracks perpendicular to the fibers. This initial damage mode requires that the strain-to-failure of the brittle fiber exceeds that of the matrix and that the interfacial deboning occurs readily. As the load is increased, the matrix cracks traverse the composite cross section with the bridging fibers remaining intact. Further loading results in significant load transfer to the intact fibers at the crack plane and subsequent interfacial fracture and pullout of these fibers from the matrix. The pullout process significantly increases the elongation of these materials prior to critical failure. Critical composite failure results when the strength of the bridging fibers is exceeded.

R. A. Shimansky, Graduate Student; H. T. Hahn, Professor; Composites Manufacturing Technology Center, Department of Engineering Science and Mechanics, The Pennsylvania State University, University Park, PA 16801.

The matrix cracking stress that initiates subcritical damage is of primary importance to the design of these composites since large, surface-connected matrix cracks will expose the delicate fibers to corrosive and oxidative environments. Recent modeling efforts, using both discrete[6,7] and continuous[8,9] composite geometries, have predicted the composite stress required for matrix cracking. While these modeling approaches are fundamentally sound, they assume a nominal interfacial shear strength that does not explicitly include interfacial debonding and friction properties. Therefore, these models cannot evaluate the effect interface properties have on matrix cracking stress. The purpose of this work is to develop a scheme that locally characterizes the fiber pullout response and integrates this local response into a global fracture model that predicts matrix cracking stress. The advantages of this scheme over previous models lie in its ability to estimate the effect of interfacial properties such as fiber/matrix mismatch, friction, and debond strength on matrix cracking stress.

ANALYSIS

Previous Work

A global model for matrix cracking, based upon a fracture mechanics approach, has been derived by Marshall, Cox and Evans[8] (hereto referred as MCE). In this work, a uniaxially stressed homogeneous isotropic composite plate is used, where intact fibers bridge a central matrix crack of length 2c oriented perpendicular to the fibers. To evaluate the shielding effect of bridging fibers, MCE employed a simple shear-lag analysis to determine the

relationship between fiber axial traction, σ, and fiber/matrix displacement, u. Under the assumption that a constant shear stress is developed at the interface, and that pullout occurs when this shear stress reaches a critical value, the fiber stress-displacement relation along the crack face is obtained from a simple shear lag analysis as [8]

$$\sigma(X) = \left[\frac{4\,\tau\,E_f}{r_f} \left(1 + \frac{c_f E_f}{c_m E_m} \right) u(X) \right]^{\frac{1}{2}}$$

(1)

where

$\sigma(X)$	fiber axial stress
$u(X)$	fiber/matrix pullout displacement
τ	interfacial shear strength
r_f	fiber radius
c_f, c_m	volume fraction of fiber, matrix
E_f, E_m	elastic modulus of fiber, matrix
X	normalized coordinate in crack direction = x/c
c	matrix crack length

The composite stress p(X) normal to the matrix crack surface may be assumed equivalent to $c_f \sigma(X)$, and the resulting composite stress intensity factor, K_I^*, as a function of composite stress, S_∞^*, is given by

$$K_I^* = \int_0^1 \frac{\left[S_\infty^* - p\{ u(X) \} \right]}{\sqrt{1 - X^2}}\, dX$$

(2)

Further, the MCE model assumes that , for short cracks, crack opening displacement u(X) follows the profile of an unshielded crack as obtained from

$$u(X) = \frac{2\left(1 - v^{*2}\right)}{E^*} K_I^* \left(\frac{c}{\pi}\right)^{\frac{1}{2}} \left(1 - X^2\right)^{\frac{1}{2}}$$

(3)

To determine the matrix cracking stress, a fracture criterion based upon the critical stress intensity of the matrix is employed,

$$K_{IC}^m = \frac{E_m}{E^*} K_I^* = K_{IC}^m \qquad \text{when } S_\infty^* = S_{mc}^*$$

(4)

where

K_{IC}^m	fracture toughness of matrix
S_∞^*	composite uniaxial tensile stress, parallel to fibers
S_{mc}^*	matrix cracking stress
E^*	major elastic modulus of composite = $c_f E_f + c_f E_m$
v^*	major Poisson's ratio of composite
v_f, v_m	Poisson's ratio of fiber, matrix

To determine matrix cracking stress for short cracks, Eq. 2 is solved using the fracture criterion of Eq. 4 and the crack face stress distribution of Eqs. 1 and 3. Although this analysis provides insight on how interfacial strength influences matrix cracking stress, it does not delineate the individual effects that thermal expansion mismatch, friction and debonding have on this interfacial strength.

Recently, more rigorous analytical studies of the fiber pullout problem shown in Figure 1 include Poisson contraction, residual stresses normal to the interface as a result of thermal expansion mismatch, and frictional[10], chemical[11] and frictional-chemical[12] bonding at the interface. These pullout models show that fiber pullout behavior may be radically different from that predicted by MCE's fiber pullout (Eq. 1). By including the Poisson's contraction of the fiber during pullout, the improved pullout model shows a more compliant fiber pullout behavior than that predicted by MCE's shear-lag analysis.

To evaluate the effects of interfacial properties on matrix cracking stress, improved pullout models must be employed. The problem with using these enhanced pullout analyses lies in obtaining a closed-form pullout function and integration of this function to determine crack tip stress intensity. The objective of this work is to incorporate a more realistic fiber pullout analysis into the MCE matrix cracking model so that the effect of different interfacial properties on matrix cracking stress may be studied.

Present Work

To determine the pullout response of bridging fibers, a heterogeneous single-fiber model was employed, as shown in Figure 1. The fiber pullout response for this local model geometry has been analyzed by Gao et al.[12]. This model assumes that fiber pullout is controlled by propagation of a frictionally shielded Mode II interface crack. It includes the effect of radial stresses at the interface due to constituent thermal and shrinkage mismatch. The pullout response was determined numerically by an incremental debonding/sliding scheme. For brevity the exact equations governing pullout of a fiber are not provided, however, the functional form of these relations is given by

$$\sigma(X) = F\left(C_f, C_m, c_f, q_o, \mu, \eta, u(X)\right)$$

(5)

where

C_f, C_m	elastic properties of fiber and matrix
q_o	radial mismatch stress normal to interface

μ interface friction coefficient

η interface debond energy

From this fiber pullout model the effects of thermal mismatch, interface friction coefficient and interface debonding energy are schematically illustrated in Figures 2a-2c. Due to the complexity of Eq. 5, an iterative procedure is required to obtain the pullout response of Figure 2. Obtaining a simple closed form solution similar to Eq. 1 is not possible. However, the following pullout model was used to approximate the pullout data of Figure 2. Linear regression analysis was performed to determine the parameters U and a.

$$\sigma(X) = \left(\tilde{\sigma} - \sigma_d\right)\left[1 - \exp\left\{-\left(\frac{u(X)}{U}\right)^a\right\}\right] + \sigma_d \tag{6}$$

where

$\tilde{\sigma}$ fiber axial stress to nullify radial interference of fiber and matrix

σ_d fiber axial stress to initiate interface debonding (see Figure 2b)

Using the pullout response of Eq 6, where $p(X) = c_f \sigma(X)$, and the crack opening profile of Eq. 2, we now determine $p(u\{X\})$ as

$$p(X) = c_f \sigma(X) = c_f\left[\tilde{\sigma} - \left(\tilde{\sigma} - \sigma_d\right)\exp\left\{\left(\frac{\phi K_I^*}{U}\right)^a \left(\frac{c}{\pi}\left(1-X^2\right)\right)^{\frac{a}{2}}\right\}\right] \tag{7}$$

where

ϕ $= 2\left(1 - \nu^{*2}\right) / E^*$

Solving the crack tip stress intensity integral of Eq. 2 with Eq. 7 and substituting the fracture criterion of Eq. 4 gives the matrix cracking stress as

$$S_{mc}^* = \frac{K_{IC}^*}{\sqrt{\pi c}} + c_f\tilde{\sigma} - \frac{I(a)}{\sqrt{\pi}} c_f\left(\tilde{\sigma} - \sigma_d\right)\exp\left\{-\left(\frac{\phi K_{IC}^*}{U}\right)^a \left(\frac{c}{\pi}\right)^{\frac{a}{2}}\right\} \tag{8}$$

where

$$I(a) = \int_0^1 \frac{\exp\left[\left(1-X^2\right)^{\frac{a}{2}}\right]}{\sqrt{1-X^2}}\, dX = \sum_{n=0}^{\infty} \frac{1}{n!}\frac{\Gamma\left(\frac{na+1}{2}\right)}{\Gamma\left(\frac{na+2}{2}\right)}$$

Note that Eq. 8 is analogous to Eq. 21 of Ref. 8. Eq. 8 provides an approximation to the composite stress for matrix cracking. Similar to the analysis in the literature, the predicted S_∞^* does not monotonically decrease with increasing crack length. Thus, the minimum stress is taken as the matrix cracking stress. The corresponding crack length at the minimum matrix cracking stress defines the transition crack length. This transition value defines the upper limit of crack length for which Eq. 8 remains valid.

RESULTS AND DISCUSSION

In this study, matrix cracking stress was numerically evaluated for two fiber-reinforced ceramic matrix composites based on Eqs. 6 and 8 . The two materials were selected because of the ready availabilty of data that includes room temperature matrix cracking stress.

Friction Bonded Interface

The first material, developed by Corning Glass, consists of 35% volume fraction Nicalon™ SiC fibers embedded in a calcium-aluminosilicate (CAS) glass-ceramic matrix. The fiber/matrix load transfer is accomplished by constituent mismatch from thermal shrinkage (from processing temperature of 1250°C) and friction between the fiber and matrix surfaces. The constituent and composite mechanical properties for this system are given in Table 1.

To evaluate the effect of thermally induced residual stress, the operating temperature was varied to develop a range of mismatch stresses normal to the interface. Using the local fiber pullout model from Eq 6, we determined the pullout response for service temperatures from 100 to 1000°C, with the friction coefficient set to a constant value of 0.25. Using Eq. 8, the matrix cracking stress was determined over the range of service temperatures. Similarly, to assess the effect of friction, the pullout response and matrix cracking stress were determined for a range of friction coefficients from 0.1 to 0.6, at room temperature of 20°C. The range of friction coefficients was based upon a similar values obtained by fiber pushout[13] and pullout tests [14] for a similar glass matrix composites. Data obtained by these analyses are presented in Table 2.

As the operating temperature is increased, the thermal mismatch stresses q_o at the interface are lowered. This lowering of the normal stress reduces the frictional strength along the interface, facilitating slip of the bridging fibers and a corresponding decrease in matrix cracking stress. This predicted performance agrees with elevated tensile test data obtained by Mandell et al. for a Nicalon/1723 composite[15]. At lower service temperatures the model predicts a slight increase in matrix cracking stress. We attribute this to small inaccuracies in the regression analysis used to determine the parameters a and U. However, as the service temperature approached the stress-free (processing) temperature, a significant decrease in matrix cracking stress is predicted. This behavior is shown in Figure 3.

In terms of friction , Eq. 8 predicts the experimentally determined matrix cracking stress of 200 MPa reasonably well for this composite using friction coefficient of about 0.25. The predicted results exhibit the expected trend that increases in friction between the fiber and matrix will increase matrix cracking stress, especially at lower friction coefficient values like those measured experimentally. In Figure 4, the relationship between matrix cracking stress and friction coefficient is shown.

Table 1. Constituent and composite properties for Nicalon/CAS.

Mechanical Property	Nicalon® SiC fiber	CAS matrix	composite $c_f = 0.35$ [16]
ultimate strength, σ_u (GPa)	2.1	unk	0.444
ultimate strain, ε_u (%)	0.01	unk	0.1-0.16
long. modulus, E_{11} (GPa)	200	88	124
trans. modulus, E_{22} (GPa)	- - -	- - -	117
Poisson's ratio	0.29	0.22	0.24
longitudinal CTE ($\mu\varepsilon$/°C)	3.9	5.0	4.3
transverse CTE ($\mu\varepsilon$/°C)	unk.	- - -	4.5
Other Properties	fiber dia. 16 μm	fracture toughness ~2 MPa√m	matrix crack stress 200 MPa

Table 2. Numerical data obtained from analyses of Nicalon/CAS.

μ	η/η_m	T (°C)	q_0 (MPa)	$\tilde{\sigma}$ (MPa)	σ_d (MPa)	a	U (μm)	S_{mc}(MPa)
0.10	0	20	59.9	1226	0	0.525	0.817	22.90
0.15	"	"	"	"	"	0.532	0.525	95.10
0.20	"	"	"	"	"	0.538	0.384	150.4
0.25	"	"	"	"	"	0.544	0.301	195.4
0.30	"	"	"	"	"	0.549	0.247	233.4
0.35	"	"	"	"	"	0.555	0.209	266.3
0.40	"	"	"	"	"	0.560	0.181	295.0
0.50	"	"	"	"	"	0.570	0.142	343.4
0.60	"	"	"	"	"	0.579	0.117	382.6
0.25	"	200	51.2	1047	"	0.549	.253	221.0
"	"	400	41.4	847.2	"	0.556	0.201	238.5
"	"	600	31.7	647.3	"	0.567	0.151	236.9
"	"	800	21.9	448.2	"	0.584	0.102	199.2
"	"	1000	12.2	249.2	"	0.621	0.056	109.9

Friction-Chemical Bonded Interface

The second material, developed by NASA[18], consists of 30% volume fraction of AVCO SCS-6 ß-SiC fibers surrounded by reaction-bonded Si_3N_4. During processing the stress-free composite temperature is 1400°C. The constituent and composite mechanical properties of this system are given in Table 3. In this system, two different interface conditions were assumed. Interface A assumes the same frictional bonding behavior as that for the Nicalon/CAS composite discussed above. Interface B assumes the existence of both frictional and chemical bonds. For this interface, a friction coefficient of 0.30 was held constant, while the interface debond energy was varied from 0 to 4% of the matrix debond energy. Debond energies in excess of 4% of that of the matrix (for this composite only) require fiber stresses that eliminate interference contact due to Poisson's contraction of the fiber loaded in tension. When this occurs all composite load is carried by the intact fibers, increasing pullout displacement significantly reducing the ability of fibers to shield the matrix crack. The result is a matrix crack that rapidly traverses the matrix.

Using the local fiber pullout model, the fiber stress-displacement relations were established for both interfaces A and B. The matrix cracking stress was determined as a function of interface friction, as shown in Figure 5. Matrix cracking stress as a function interface debond energy for interface B is shown in Figure 6. Input data required by Eq. 8 for each interface is given in Table 4.

For interface A, the experimental matrix cracking stress is predicted when the friction coefficient of about 0.50 is used. Based on the analytical results, it appears that chemical bonding does not play a more significant role on the matrix cracking behavior of this composite.

For interface B, the experimental matrix cracking stress is predicted when the interface debond energy is about 4% of the matrix value of 138 N/m. This relatively low debond energy is reasonable for this composite, where processing reduces the chemical bonding between fiber and matrix. As the interface debond energy is increased, the predicted matrix cracking stress shows a corresponding increase. This result is expected, since the fiber traction, σ (proportional to the composite stress $S_\infty{}^*$) must exceed σ_d before the matrix crack can open. Conceivably, the interface debond energy should be set such that the fiber stress for debonding, σ_d, is just below the fiber strength, σ_{fu}. This will maximize matrix cracking stress while maintaining the integrity of the bridging fibers.

Table 3. Constituent and composite properties for SCS-6/RBSN.

Mechanical Property	AVCO SiC fiber	RBSN matrix	composite $c_f = 0.30$ [17]
ultimate strength, σ_u (GPa)	3.4	0.084	0.682
ultimate strain, ε_u (%)	1.8	0.08	0.45
long. modulus, E_{11} (GPa)	400	110	193
trans. modulus, E_{22} (GPa)	unk.	n/a	69
Poisson's ratio	0.22	0.22	0.21
longitudinal CTE ($\mu\varepsilon$/°C)	4.3	4.5	4.3
transverse CTE ($\mu\varepsilon$/°C)	unk.	n/a	4.5
Other Properties	fiber dia. 142 μm	fracture toughness 4 MPa$\sqrt{m}$	matrix crack stress 227 MPa

Table 4. Numerical data obtained from analyses of SCS-6/RBSN.

μ	η/η_m	T (°C)	q_0 (MPa)	$\tilde{\sigma}$ (MPa)	σ_d (MPa)	a	U (μm)	S_{mc}(MPa)
0.10	0	20	16.7	631	0	0.545	2.719	86.8
0.15	"	"	"	"	"	0.565	1.741	125.5
0.20	"	"	"	"	"	0.578	1.273	153.9
0.30	"	"	"	"	"	0.601	0.828	192.7
0.40	"	"	"	"	"	0.620	0.618	216.0
0.50	"	"	"	"	"	0.637	0.497	227.4
0.60	"	"	"	"	"	0.652	0.419	231.2
0.70	"	"	"	"	"	0.666	0.354	232.1
0.30	0.005	"	"	"	137.5	0.669	1.102	196.6
"	0.010	"	"	"	193.0	0.648	1.055	204.3
"	0.015	"	"	"	235.6	0.622	0.966	210.4
"	0.020	"	"	"	271.6	0.591	0.848	215.2
"	0.025	"	"	"	303.2	0.555	0.705	219.1
"	0.030	"	"	"	331.8	0.513	0.542	222.4
"	0.035	"	"	"	358.0	0.466	0.367	225.2
"	0.040	"	"	"	382.7	0.406	0.190	227.9

CONCLUSIONS

In this study, an existing model to predict matrix cracking stress in brittle matrix composites was improved by utilizing a more realistic fiber pullout behavior. The result is model that predicts matrix cracking stress as a function of fiber/matrix thermal mismatch, interfacial friction coefficient and interfacial debond energy. The model was used to predict matrix cracking stress for two different brittle matrix composite systems for a range of assumed interfacial conditions. The predictions of the model concur with experimental data when resonable interfacial properties are used. This approximate model qualitatively shows that matrix cracking stress will increase with interface friction coefficient and debond energy, and decrease at high service temperature.

Future work includes using the pullout analysis to evaluate fiber pushout test results that experimentally determine the interfacial properties. Once the interfacial parameters are known, accuracy of the matrix cracking model can be evaluated. Such a comprehensive evaluation should include a single composite system processed such that interfacial properties are varied. For the case of chemically bonded fibers, the crack opening profile of Eq. 3 may be invalid at the microscopic level. Fibers near the crack tip that are not debonded will cause crack closure, significantly changing the crack opening profile.

ACKNOWLEDGEMENTS

This work was supported by the Center for Advanced Materials of the Pennsylvania State University under its Cooperative Program in High Temperature Materials.

REFERENCES

1. Phillips, D.C., "Interfacial Bonding and the Toughness of Carbon Fiber Reinforced Glass and Glass Ceramics" J. Mater. Sci., Vol. 9, Nov. 1974, pp. 1847-54.

2. Phillips, D.C., "The Fracture Energy of Carbon-Fiber Reinforced Glass" J. Mater. Sci., Vol. 7, Oct. 1972, pp. 1175-91.

3. Prewo, K.M. and Brennan, J.J., "High Strength Silicon Carbide Fiber Reinforced Glass Matrix Composites," J. Mater. Sci., Vol. 15, Feb. 1980, pp. 463-468.

4. Prewo, K.M. and Brennan, J.J., "Silicon Carbide Yarn Reinforced Glass Matrix Composites," J. Mater. Sci., Vol. 17, Apr. 1982, pp. 1201-06.

5. Brennan, J.J. and Prewo, K.M., "Silicon Carbide Fiber Reinforced Glass-Ceramic Matrix Composites Exhibiting High Strength and Toughness," J. Mater. Sci., Vol. 17, Aug. 1982, pp. 2371-83.

6. Aveston, J., Cooper, G.A. and Kelley, "The Properties of Fiber Composites," Conference Proceedings, National Physical Laboratory, Guildford. IPC Science and Technology Press Ltd. , 1971, pp 15-26.

7. Budiansky, B., Hutchinson, J.W. and Evans, A.G., "Matrix Fracture in Fiber-Reinforced Ceramics," J. Mech. Phys. Solids, Vol. 34, Feb. 1986, pp 167-189.

8. Marshall, D.B., Cox, B.N. and Evans, A.G., "The Mechanics of Matrix Cracking in Brittle-Matrix Fiber Composites," Acta. Metall. Vol. 33, Nov. 1985, pp 2013-21.

9. Marshall, D.B, and Evans, A.G., "Tensile Failure of Brittle Matrix Composites," Fifth International Conference on Composite Materials, San Diego, CA , 1985.

10. Dollar, A. and Steif, P.S., "Load Transfer in Composites with a Coulomb Friction Interface," submitted Int. J. Solids and Struct. (1988).

11. Stang, W. and Shah, S.P., "Failure of Fiber-reinforced Composites by Fiber Pullout Fracture," J. Mat. Sci., Vol. 21, 1986, pp 953-971.

12. Gao, Y.-C., Mai, Y.-W. and Cotterell, B., "Fracture of Fiber Reinforced Materials," J. of App. Math Phys. (ZAMP) , Vol. 39, 1988, 550-572.

13. Faber, K.T., Advani, S.H., Lee, J.K. and Jinn, J.-T., "Frictional Stress Evaluation Along the Fiber-Matrix Interface in Ceramic Matrix Composites," J. Am. Ceram. Soc., Vol. 69, Sept. 1986, pp C-208-C-209.

14. Griffin, C.W., Limaye, S.Y., Richardson, D.W. and Shetty, D.K., "Evaluation of Interfacial Properties in Borosilicate-SiC Composites Using Pullout Tests," Ceram. Eng. Sci. Proc., Vol. 9, 1988, pp. 671-678.

15. Mandell, J.F., Grande, D.H. and Edwards, B., "Test Method Development for Structural Characterization of Fiber Composites at High Temperatures," Ceram. Eng. Sci. Proc., Vol. 7, 1986, pp. 524-535.

16. "Nicalon/Glass-Ceramic: Properties of CGW Nicalon/CAS II Glass-Ceramic Composites" Information sheet from Corning Glass, Corning, NY (1988).

17. Bhatt, R.T. and Phillips, R.E., "Laminate Behavior for SiC Fiber-Reinforced Reaction-Bonded Silicon Nitride Matrix Composites," NASA TM 101350, October 1988.

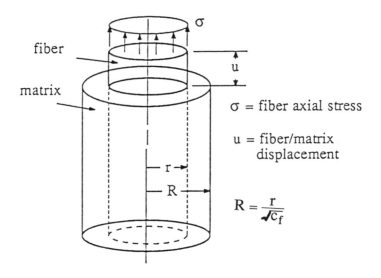

Figure 1. Representative volume element for fiber pullout [12]

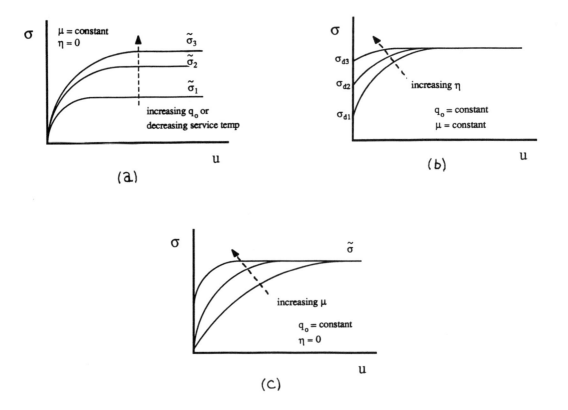

Figure 2. Effect of interfacial properties on fiber pullout response

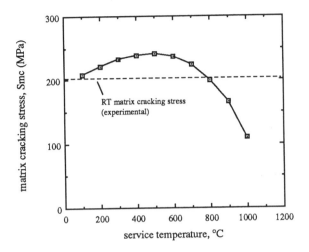

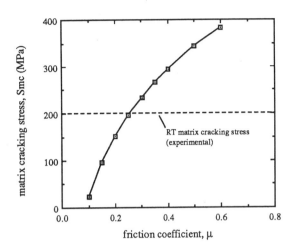

Figure 3. Predicted effect of service temperature on matrix cracking stress of Nicalon/CAS

($\sigma_d = 0$, $\mu = 0.25$).

Figure 4. Predicted effect of friction on matrix cracking stress of Nicalon/CAS

($\sigma_d = 0$, T = 20°C).

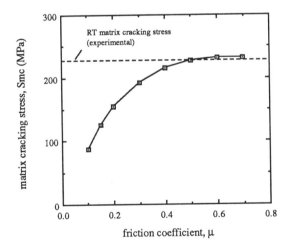

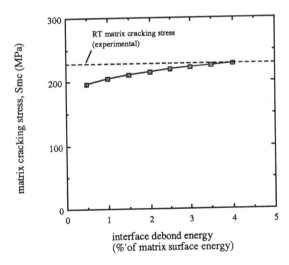

Figure 5. Predicted effect of friction on matrix cracking stress of AVCO SCS-6/RBSN

($\sigma_d = 0$, T = 20°C).

Figure 6. Predicted effect of debond energy on matrix cracking stress of AVCO SCS-6/RBSN
($\mu = 0.3$, T = 20°C).

Microstructural Characterization of Ceramic Matrix Composite Fiber Reinforcement Using Nuclear Magnetic Resonance Spectrometry

ROBERT A. MARRA AND **NEAL R. DANDO**

ABSTRACT

High resolution solid state magic angle spinning nuclear magnetic resonance spectrometry is capable of unambiguously and nondestructively characterizing the ceramic fiber reinforcement component of intact ceramic composite parts. Structure/property relationships are elucidated as a function of post processing thermal treatments of Nextel® and Nicalon® reinforced ceramic/ceramic composites.

INTRODUCTION

The local structure of amorphous oxide and nonoxide ceramics is an area of notable research emphasis in modern materials science. Ceramic composites, specifically ceramic fiber reinforced composites, are receiving considerable attention due to their potential use in high temperature structural applications [1,2,3]. Concurrent with the development of divergent processing technologies is the escalating need for improved structural characterization techniques to establish coherent chemical structure/physical property relationships in these systems.

Magic angle spinning solid state nuclear magnetic resonance spectrometry (MAS-NMR) has emerged as a powerful, non-destructive tool for characterizing inorganic solids [4,5]. The technique is ideal for investigating the coordination and chemical bonding of NMR active nuclei as well as elucidating phase distributions regardless of crystallite size or degree of crystallinity. The quantitative, multi-elemental information obtainable from MAS-NMR investigation greatly augments the physical structure information available from other methods of analysis.

The present investigation explores the capability of ^{11}B, ^{13}C, ^{27}Al and ^{29}Si MAS-NMR for nondestructively characterizing the thermal

Alcoa Laboratories, Alcoa Center, Pennsylvania 15069

alumina-based ceramic matrix composites(CMC). The goal of this work is to unambiguously identify temperature dependent chemical bonding and phase changes in the fiber reinforcement of composite parts as a function of post-processing conditions.

EXPERIMENTAL

Both Nicalon® SiC (Nippon Carbon Company, Tokyo, Japan) and Nextel® 440 alumina-silica (3M Company, St. Paul, Minnesota) continuous ceramic fibers have been examined as reinforcement for an Alcoa proprietary alumina-based ceramic matrix composite. The composites were fabricated by a slurry infiltration process [6] in which woven fabric of the continuous ceramic fiber is infiltrated with an alumina based slurry. The infiltrated fabric plies are laid-up on a mold using techniques similar to those used in organic matrix composite fabrication. The "green" composite is cured at low temperatures and pressures (similar to those obtainable in conventional autoclaves) to produce the final near net shaped component. The as-produced composite contains approximately 50 weight % fiber in a matrix containing about 10-15 % total porosity.

MAS-NMR analyses were performed using a GE GN-300 wide bore instrument with a 7.05T cryomagnet and Chemagnetics solids accessories. ^{29}Si spectra were acquired at 59.6 MHz using single pulse excitation with high power proton decoupling. Samples were placed in Delrin rotors and spun at a sample rotation rate of 3kHz. ^{27}Al and ^{11}B spectra were acquired at 78.2 and 96.3 MHz, respectively, using single pulse, 1 usecond excitation[approximatedly 30 degrees, measured on $Al(H_2O)_6$]. The samples were placed in Torlon rotors and spun at rotation rates of 8kHz. The spectra are referenced to the known chemical shifts of $Al(H_2O)_6$ or BPO_4, resptively.

The as-received fibers, as-produced composites, and heat treated composites were characterized by ^{11}B, ^{12}C, ^{27}Al, and ^{29}Si MAS-NMR. The heat treatments were preformed in air at temperatures of 800°C, 1300°C, and 1540°C for the Nicalon® fiber composite and 800°C and 1300°C for the Nextel® fiber material. The exposure time at the elevated temperature was typically less than 15 minutes.

RESULTS AND DISCUSSION

Nicalon® is one of the most widely used ceramic fibers for CMC studies. The fiber is composed of 1.7 to 2.5 nm beta SiC crystallites in an amorphous matrix which contains a substantial amount of excess carbon and oxygen [7]. The ^{29}Si MAS-NMR spectrum of a Nicalon® fiber reinforced CMC sample is shown in Figure 1. All of the signal intensity observed arises from ^{29}Si nuclei present in the fiber reinforcement, since the matrix is devoid of Si. The spectrum consists of a broad resonance with a center of mass at -18 ppm and low intensity shoulders centered about -35, -70 and -110 ppm. The overall spectrum is little different from that of the neat fibers, which has been reported previously [8]. Comparison of Figure 1 with published NMR investigations of SiC indicate that the broad resonance encompasses the entire chemical shift range encountered for beta and alpha SiC [8,9,10]. The shoulders are evidence of small amounts of elemental silicon and silica [7]. While the center of mass of the main resonance in Figure 1 is consistent with the known chemical shift of SiC, the width of this resonance indicates a lack of long range order in this system. This point is illustrated in Figure 2 A-C, which show the ^{29}Si MAS-NMR spectra of as received Ibiden® SiC powder(2A), a sintered part made from the powder(2C) and a powder produced by grinding the sintered sample to the particle size equivalent to that of the original powder(2B). The original powder clearly exhibits a spectrum similar to that in Figure 1. Upon sintering(at 2100°C), the development of long range order gives rise to significantly narrowed ^{29}Si resonances, indicative of beta and alpha (6H and 15R polytypes) SiC. NMR is unaffected by changes in long range order on a macroscopic level, as shown by the resolution remaining, even after grinding the sintered part(Figure 2B). While the sintering temperature was high enough to induce beta-alpha transformation, comparison all of the spectra in Figure 2 still suggests the existence of amorphous alpha SiC in the starting powder. This observation would not be detected by characterization methodologies dependent on the development of long range order.

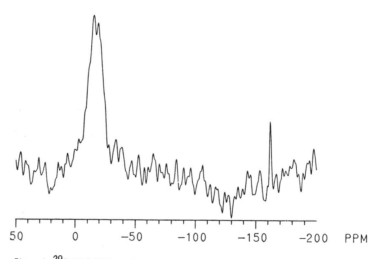

Figure 1. ^{29}Si MAS-NMR spectrum of Nicalon® fiber reinforced composite.

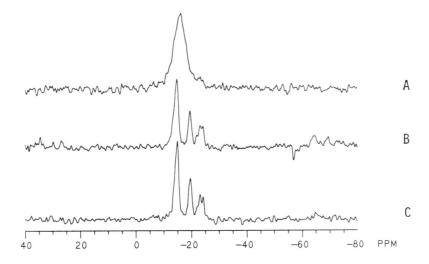

Figure 2. ^{29}Si MAS-NMR spectra of A) Ibiden® powder, B) crushed powder from a sintered part made from Ibiden® powder and C) a piece of sintered ceramic made from Ibiden® powder.

Mechanical characterization of thermal stability [1,2] may be augmented by the elucidation of temperature dependent chemical phenomena. Figures 3A-C show the ^{29}Si MAS-NMR spectra of Nicalon® reinforced CMC samples which were thermally treated(post processing) to 800°(3A), 1300°(3B) and 1540°C(3C). One clearly observes a gradual narrowing of the resonance centered at -18 ppm and the growth of a resonance centered at -115 ppm. The narrowing suggests that some development of longer range order occurs(e.g. crystal growth), while the resonance at -115 ppm shows the gradual transformation of Si to Q4 silica species within the reinforcement fibers or the development of a glassy interfacial phase. This phenomenon undoubtedly contributes to the degradation of mechanical properties observed in Nicalon® reinforced materials at high temperature [1].

Nextel® 440 is an oxide fiber produced by a sol-gel process, and is comprised of Al_2O_3, SiO_2 and small amounts of B_2O_3, added as a processing and densification aid and to inhibit grain growth. Figures 4A-C are the ^{29}Si, ^{27}Al and ^{11}B MAS-NMR spectra of neat fiber, respectively. The ^{29}Si spectrum is appreciably narrowed, indicating that the Si largely exists as crystalline material in as received fibers. The chemical shift observed agrees with reported values [11] for Q4(Al2-3) species in mullite [5], consistent with the relatively high Al/Si ratio observed in these fibers. The Al spectrum suggests the presence of at least three types of Al species, suggesting that Al is not homogeneously distributed. The large asymmetric resonance at 0 ppm is indicative of octahedral Al species, similar to what is observed in alumina oxide. The 80-20 ppm region in Figure 4B indicates the

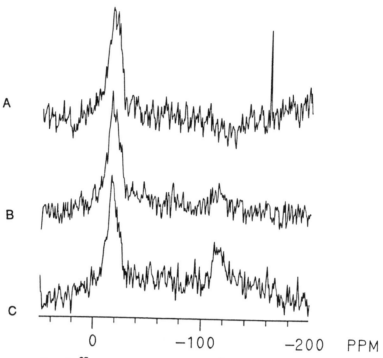

Figure 3. ^{29}Si MAS–NMR spectra of a Nicalon® reinforced CMC sample which was thermally treated at A) 800C, B) 1300C and C) 1540C.

presence of at least two resonances, owing to tetrahedral Al, centered at 58 and 48 ppm, respectively. The 58 ppm resonance is consistent with Al incorporated into either a silica or alumina lattice framework [11]. The resonance centered at 49 may arise due to BO_3 and/or BO_4 incorporation into the second coordination sphere of Al [4]. The information presented here is insufficient to unambiguously assign the source of these resonances.

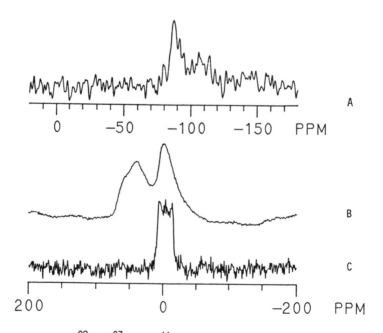

Figure 4. A) ^{29}Si, B) ^{27}Al and C) ^{11}B MAS–NMR spectra of Nextel® fiber.

While Nextel® fiber contains only 2% B_2O_3, a good signal to noise ratio ^{11}B MAS-NMR spectra may be acquired in a short period of time, owing to the reasonable relative sensitivity of [11]B and the short acquisition cycle times employed. The ^{11}B NMR spectrum in Figure 4C exhibits two resonances, a single peak resulting from tetrahedral boron at -15 ppm, and a complex, two peaked quadrupolar lineshape, with peak maxima at 0 and -35 ppm which arises from trigonal boron. [12]. The chemical shift information presented here is insufficient to unambiguously assign the second coordination sphere chemistry of the species observed.

Figures 5A and B show the ^{11}B MAS-NMR spectra of Nextel® fiber reinforced CMC samples which were heated to 800° and 1300°C, respectively. The higher temperature treated material clearly indicates that a gradual transition of trigonal boron to tetrahedral boron has occurred, evidenced by the increase in the resonance at -15 ppm. The narrowing effect of thermal treatment on the trigonal boron resonance also supports a model in which a gradual perturbation from trigonal symmetry has occurred. The observed narrowing is due to a change in the asymmetry parameter of the electrostatic field gradient experienced by trigonal boron upon thermal treatment. The temperature dependent phenomena observed support a model in which borosilicate/aluminoborate glass formation occurs in the reinforcement or at the fiber/matrix interface at elevated temperature. [27]AL was not investigated, owing to large background signals arising from the matrix employed. Characterization of reinforcement chemistry could be augmented by investigating the thermally treated samples by ^{29}Si NMR.

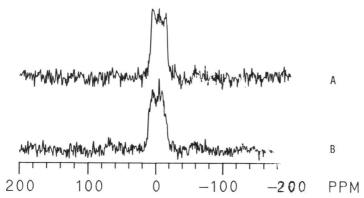

Figure 5. ^{11}B MAS-NMR spectra of Nextel® fiber reinforced CMC samples thermally treated at A)800C and B) 1300C.

SUMMARY

The data presented illustrate the potential of MAS-NMR for the non-destructive characterization of ceramic fiber reinforcements while incorporated into finished parts. While presented here for ceramic matrix composites, this methodology could easily be extended to encompass organic matrix materials. Judicious choice of probe nucleus allows selective spatial characterization of composite materials, based on their chemical composition. For example, ^{27}AL MAS-NMR would have allowed characterization of the matrix portion of the Nicalon reinforced CMC samples investigated, without interference from the reinforcement. MAS-NMR holds tremendous potential for augmenting existing inorganic structural characterization technologies and is essential for the development of chemical structure/physical property relationships in composite systems.

ACKNOWLEDGEMENTS

The authors wish to acknowledge the excellent technical assistance of Colleen Gianutsos and Douglas Weaver with preparation of CMC samples and acquisition of NMR data, respectively. The authors wish to thank the Aluminum Company of America for the permission to publish this manuscript.

REFERENCES

1. Mah, T., Hecht, N., McCullom, D., Hoenigman, J., Kim, H., Katz, A., Lipsitt, H.,"Thermal Stability of SiC Fibres(Nicalon®)," Journal of Materials Science, Vol. 19, 1984, pp. 1191-1201.

2. Simon, G., Bunsell, A., "Mechanical and Structural Characterization of the Nicalon Silicon Carbide Fiber," Journal of Materials Science, Vol. 19, 1984, pp. 3649-3657.

3. Ichikawa, HJ., Teranishi, H., Ishikawa, T., "Effect of Curing Conditions on Mechanical Properties of SiC Fibre(Nicalon®)," Journal of Materials Science Letters, Vol. 6, 1987, pp. 420-422.

4. Hahnert, M., Hallas, E., "NMR-Untersuchungen an Aluminoborat-Glasern," Revue de Chemie Minerale, Vol. 24, 1987, pp. 221-233.

5. Turner, G., Kirkpatrick, R., Risbud, S., Oldfield, E., "Multinuclear Magic-Angle Spinning Nuclear Magnetic Resonance Spectroscopic Studies of Crystalline and Amorphous Ceramic Materials," American Ceramic Society Bulletin, Vol 66, 1987, pp. 656-63.

6. Bray, D. J. , "Reinforced Alumina Composites," Paper Number EM 88-535, Fabricating Composites '88, Society of Manufacturing Engineers, 1988.

7. Sawyer, L., "Characterization of Nicalon: Strength, Structure, and Fractography," Ceramic Science and Engineering Proceedings, Vol. 6, 1985, pp. 567-575.

8. Hartman, J., Richardson, M., Sherriff, B., Winsborrow, B., "Magic Angle Spinning NMR Studies of Silicon Carbide: Polytypes, Impurities, and Highly inefficient Spin-Lattice Relaxation," Journal of the American Chemical Society, Vol. 109, 1987, pp. 6059-6067.

9. Finlay, G., Hartman, J., Richardson, M., Williams, B., "^{29}Si and ^{13}C Magic Angle Spinning NMR Spectra of Silicon Carbide Polymorphs," Journal of the Chemical Society, Chemical Communications, Vol. 3, 1985, pp159-161.

10. Guth, J., Petusky, W., "Silicon-29 Magic Angle Spinning Nuclear Magnetic Resonance Characterization of SiC Polytypes," Journal of Physical Chemistry, Vol. 91, 1987, pp. 5361-5364.

11. Mueller, D., Hoebbel, D., Gessner, W., "^{27}AL NMR Studies of Aluminosilicate Solutions, Influences of the Second Coordination Sphere on the Shielding of Aluminum," Chemical Physics Letters, Vol. 84, 1981, pp. 25-29.

12. Turner, G., Smith, K., Kirkpatrick, R., Oldfield, E., "Boron-11 Nuclear Magnetic Resonance Spectroscopic Study of Borate and Borosilicate Minerals, and a Borosilicate Glass," Journal of Magnetic Resonance, Vol. 67, 1986, pp. 544-50.

Matrix Cracking Initiation in Brittle-Matrix Composites— Experiment and Predictions

A. S. D. WANG AND MICHEL BARSOUM

ABSTRACT

This paper presents some investigative results of matrix cracking stress in unidirectional fiber-reinforced ceramic matrix composites. Several predictive models are reviewed and applied to a specific specimen loading condition (three-point bending) for which experiments are conducted using SiC/LAS (50 v %), SiC/Borosilicate (17 fiber v %) and C/Borosilicate (45 fiber v %) specimens. In each case, matrix cracks transverse to the fibers are induced in the tensile side of the specimen. A sputtered gold-film technique is used to determine the critical matrix crack stress in real-time. Results for the matrix cracking stress calculated from the mechanics models are compared with that from the experiment.

INTRODUCTION

When a unidirectional fiber-reinforced ceramic matrix composite is loaded in tension in the fiber direction, matrix cracks transverse to the fibers may be induced. Generally, this type of matrix cracking occurs if the tensile fracture strain of the matrix is lower than that of the fibers. An idealized matrix crack problem is one that forms across all the fibers, with the fibers bridging the crack. Aveston, Cooper and Kelly (ACK) [1] were the first to treat this idealized problem by using an energy balance approach. Several material, geometric and kinematic parameters are deemed to play a role in the formation of such a matrix crack: matrix crack-surface energy, interface debonding energy, frictional energy as fiber slides against the matrix, losses and/or gains in elastic strain energies as a consequence of fiber-matrix load-transfer, etc. The ACK model provides a closed form equation for the critical matrix cracking strain if the interface debonding energy is neglected:

$$(e_m)_{cr} = \{[12 \gamma_m \tau E_f (V_f)^2] / [E_c E_m^2 V_m R]\}^{1/3} \tag{1}$$

where γ_m is the fracture surface energy of the matrix, E_c is the modulus of the composite, R is the fiber diameter, V_f is the fiber volume fraction and τ is the fiber-matrix interface shear strength.

It is noted that the ACK model contains two in-situ material failure parameters. These are the matrix surface energy γ_m and the fiber-matrix interface shear strength τ. If both fof the failure parameters are known, the critical matrix cracking strain $(e_m)_{cr}$ is then directly proportional to $(\gamma_m \tau)^{1/3}$.

Budiansky, Hutchinson and Evans (BHE) [2] used a similar energy balance formulation for what they termed as steady-state-cracking. The model allows for evaluation of interface debonding and friction separately; and it includes thermal residual stress effects due to fiber-matrix thermal expansion mismatch. The BHE model predicts essentially the same matrix crack stress as the ACK model for

Department of Mechanical Engineering and Mechanics and Department of Materials Engineering, respectively, Drexel University, Philadelphia, PA 19104.

large size fiber-bridged matrix cracks.

Marshall, Cox and Evans (MCE) [3] considered the fiber-bridged matrix cracking problem from a fracture propagation stand-point. The model treats matrix cracks of small size. Mode-I crack-tip stress intensity factor K_c of the composite is evaluated by an approximate stress field analysis and it is used as the crack propagation driving force within the context of fracture mechanics. The composite stress σ_c, required to propagate a matrix crack of arbitrary length c, is calculated from the following equation:

$$(\sigma_c / \sigma_m) = (1/3)(c/c_m)^{-1/2} + (2/3)(c/c_m)^{1/4}$$

(2)

where c_m is a characteristic matrix crack size defined as:

$$c_m = (\pi K_c / \alpha I^2)^{2/3}$$

and σ_m is a characteristic stress defined as

$$\sigma_m = (3/\sqrt{\pi})(K_c^2 \alpha I^2)^{2/3}$$

while $K_c = K_m E_c E_m$, I is a dimensionless crack geometry constant (=1.2 for a straight crack), and

$$\alpha = [8(1-\nu^2) \tau V_f^2 E_f] / [E_m V_m R \pi^{1/2}]$$

with the quantity ν being the composite Poisson ratio transverse to the fibers.

The MCE model as represented by Eq. (2) provides a non-dimensional relation between σ_c/σ_m and c/c_m for small matrix cracks having initial length $c < c_m/3$. Note the connection between the parameters c_m and σ_m and the material failure parameters γ_m and τ through the quantities K_c and α.

For crack length longer than $c_m/3$, the crack is considered long; and the critical stress (or strain) for propagation is given separately by ACK, Eq. (1). Hence, it is noted that Eq. (2) does not, generally, agree with the result predicted by Eq. (1) when c/c_m is large; hence, it should not be used for large sized matrix cracks.

Recently, McCartney [4] modified the MCE model for short cracks by introducing a distributed traction on the crack surfaces to represent the bridging fibers. A characteristic matrix crack size a_0 is defined which depends on the composite stress intensity factor K_c (related to K_m or γ_m) and the interfacial shear strength τ:

$$a_0 = (\sqrt{\pi} K_c / \lambda^2)^{2/3}$$

where λ is defined as

$$\lambda = 2(V_f/V_m)(2\pi \tau E_f E_c/R E_m^2)^{1/2}$$

The critical composite stress σ_c at the propagation of the matrix crack of initial size a is given by :

$$\sigma_c / \sigma_0 = [\mu_c Y^2 (\mu_c)]^{-1/3} \qquad (3)$$

where $\sigma_0 = (K_c/\sqrt{(\pi a_0)})$, $K_c = (V_m E_c/E_m]^{1/2} K_m$ and $Y (\mu_c)$ is a complicated integral equation involving μ_c. The variable μ_c is, on the other hand, defined as

$$\mu_c = (\lambda^2 a) / (\pi \sigma_c)$$

The McCartney model, Eq. (3), also provides a universal relationship between the non-dimensional quantities σ_c/σ_0 and a/a_0. A numerical procedure is required, however, to evaluate this relationship for a specific case. The McCartney model approaches asymptotically the ACK model for long matrix cracks. It is further noted that the characteristic quantities a_0 and σ_0 in the McCartney model are not the same as c_m and σ_m in the MCE model. In fact, it has been shown that the McCartney model generally predicts a matrix cracking stress lower than that predicted by the MCE model [6] for short matrix cracks.

The BHE, MCE and McCartney models are essentially modifications of the ACK model; their descriptions of the matrix cracking behavior are essentially the same. An in-depth evaluation of these models has been given in a recent review paper by Kerans, Hay, Pagano and Parthasarathy [5].

Based on a different view point on how a matrix crack is initiated, Wang [7] presented a model which begins with the assumption that the composite inherently contains a random distribution of microflaws. Two types of the microflaws are deemed to play a major role in matrix crack initiation. These are flaws in the matrix material and flaws in the fiber-matrix interface. To model the ACK problem discussed above, for instance, Wang considers a unit-cell which represents the most probable matrix cracking site in the composite. This unit-cell is schematically shown in Figure 1, where a flaw of size a is situated in the matrix of thickness, t, between two fibers (the fiber diameter is d). In addition, a pair of interfacial flaws, each of size $2b$, is present on either side of the matrix flaw. It is then assumed that the matrix flaw of size a acts like a small crack, and that it can propagate and cause fracture of the matrix at some critically applied tension.

The introduction of the interface flaws is an indirect representation of the interfacial bonding and/or shear strength. For instance, if the interface bonding is poor then b is large; if the interface bonding is good then b is small. In particular, if the interface bonding is perfect then $b = 0$.

By a method of fracture mechanics in conjunction with a finite element procedure [8], the strain energy release rate at the tips of the matrix flaw can be calculated and expressed in the form:

$$G(e_c, \Delta T) = [\sqrt{C_e(a)}\,(e_c) + \sqrt{C_T(a)}\,\Delta T]^2\, d$$

where e_c is the applied composite tensile strain, ΔT is a temperature load residing in the composite due to fabrication, and $C_e(a)$ and $C_T(a)$ are computed coefficient functions for the energy release rate due to the application of e_c and ΔT, respectively (see [8] for details).

By a fracture mechanics criterion based on strain energy release rate, the critical composite strain at the onset of matrix cracking can be determined from:

$$G(e_c, \Delta T) = G_m \tag{4}$$

where G_m $(=2\gamma_m)$ is the critical strain energy release rate of the matrix.

It is noted that the functions $C_e(a)$ and $C_T(a)$ are keenly affected by the interfacial flaw size $2b$, and a host of other micromechanical factors, including fiber spacing, fiber diameter and fiber axial stiffness relative to the matrix's [7].

In this paper, we shall present some experimental data for matrix cracking in specimens made from several types of ceramic matrix composites and tested under 3-point bend conditions. We then apply the predictive models discussed above. In each case, we illustrate the use of the model and identify the necessary input parameters. A comparison is made between the predicted results and the experimental results.

EXPERIMENT AND RESULTS

Three unidirectional fiber-reinforced composites were tested. These are the SiC/LAS system with 50 v% Nicalon fiber and the C/Borosilicate system with 45 v% carbon fibers (all supplied by UTRC), and the SiC/Borosilicate system (with 17 v% AVCO fiber) which is fabricated in-house. The fabrication processes for the composites and the preparation details of the test samples (including the polishing of the samples) have been described elsewhere [9]. The relevant baseline properties of these systems are listed in Table 1.

TABLE 1 SUMMARY OF PROPERTIES OF COMPOSITE SYSTEMS

	SiC+/LAS@	SiC*/Borosilicate#	C**/Borosilicate#
K_m	2 MPa$\sqrt{m}$	0.75 MPa$\sqrt{m}$	0.75 MPa$\sqrt{m}$
τ	2 MPa	10 MPa	10-25 MPa
E_f	200 GPa	400 GPa	380 GPa
E_m	85 GPa	63 GPa	63 GPa
V_f	0.5	0.17	0.45
R	8 μm	70 μm	4 μm

+ Nicalon   @ Data from Ref [3]   * SCS-6 AVCO   #Corning-7740
** Hercules Type HMU

Test samples are prepared in the form of flat bars to be tested in 3-point bend. The size of the test samples varies, but is roughly 3x0.5x0.18 cm for the SiC/LAS, 3x0.5x0.2 cm for the SiC/Borosilicate and 3x0.5x0.44 for the C/Borosilicate. The lower support pin span is 2.54 cm. All tests are conducted in room temperature.

In order to determine the matrix cracking stress, the tensile side of the beam is sputtered with a thin film of gold between two parallel line-electrodes. The electric resistance of the gold film is measured before, during and after loading on the beam. At zero load, the resistance of the gold film is measured as R_0, while under loading the resistance is measured as R_{max}. The change in resistance is therefore $\Delta R_{max} = R_{max} - R_0$. Similarly, when the beam is unloaded a residual resistance R_{res} is measured. Then, the quantity $\Delta R_{res} = R_{res} - R_0$ is also recoreded which represents possible permanent damage in the beam.

Fig. 2 shows the summary experimental results for the three types of samples. Here, the quantity $(\Delta R_{max} - \Delta R_{res})/R_0$ is plotted against the maximum apparent tensile stress in the beam (the apparent tensile stress is calculated based on the simple beam theory).

It is seen from Figure 2 that the change in electric resistance in the gold film is small initially and is linearly increasing with the applied stress. However, at some critical stress the change in resistance suddenly increases, signifying the onset of matrix cracking normal to the direction of the electric current. The critical stresses for matrix cracking measured from four SiC/LAS specimens averaged approximately 375 ± 10 MPa. Similarly, the measured average matrix cracking stresses for the SiC/Borosilicate and C/Borosilicate systems are 78 ±10 MPa and 360 ±2 MPa, respectively.

PREDICTED MATRIX CRACKING STRESSES

To facilitate a prediction for the onset matrix cracking in the test specimen, we shall use the models outlined in the previous section. If we assume that the observed matrix cracking is initiated by a long matrix flaw in the context of the ACK model, a prediction can be made using Eq. (1). The relevant parameters needed in Eq. (1) for each of the material systems are listed in Table 1.

As was mentioned earlier, there are several material and geometrical parameters that appear in the ACK model; each can influence the predicted results. In particular, we mention the value of τ, the interface shear strength. Depending on the test method used and the assumption made in deducing the test data, τ can have a wide range of values. Based on the τ values given in Table 1 for each material system considered, the predicted matrix cracking stresses from the ACK model are given in the following:

SiC/LAS	SiC/Borosilicate	C/Borosilicate
265 MPa (τ = 2 MPa)	71.4 MPa (τ = 10 MPa)	572 MPa (τ = 10 MPa)

Comparing the above with the experimental results, it is seen that the ACK model predicted well the onset stress for the SiC/Borosilicate system, under predicted the stress for the SiC/LAS system and over predicted the stress for the C/Borosilicate system. But in view of the uncertainty in the definition and determination of τ, and the sensitivity of the model to a number of other parameters, it would be difficult if not inappropriate to appraise the adequacy of the model based on the limited amount of test data presented here.

The MCE and the McCartney models provides a prediction for matrix flaws of small initial size. To facilitate a prediction, one must evaluate first the characteristic quantities c_m and σ_m for MCE and a_o and σ_o for McCartney. These are calculated as follows based on the material parameters listed in Table 1 (τ = 10 MPa for the C/Borosilicate system):

	SiC/LAS	SiC/Borosilicate	C/Borosilicate
c_m	325 μm	817 μm	37 μm
σ_m	314 MPa	84 MPa	679 MPa
a_o	26 μm	94 μm	1.6 μm
σ_o	203 MPa	55 MPa	440 MPa

Note that in the MCE model, a matrix crack is considered long if $c > c_m/3$; while in the McCartney model, a long crack is when $a > 5a_o$ or so.

Thus, if we assume that onset of matrix cracking is due to propagation of a short crack in the context of MCE or McCartney, we must determine the initial matrix crack size, c or a in the respective models. Since both quantities are unavailable for any of the test cases, it is therefore not possible to make a definitive prediction using either the MCE model or the McCartney model for the present test cases.

Finally, we follow the Wang model. Referring to the sketch in Figure 1, we assume that the cause of matrix cracking is due to the propagation of the matrix flaw of size a, and that he material condition governing the propagation is the balance of the energy release rates at the matrix flaw tip, given by Eq. (4). Calculation of the strain energy release rates are then carried out by a finite element procedure.

Figures 3 and 4 show the $C_e(a)$ and $C_T(a)$ coefficients for the SiC/LAS system, respectively. In calculating these curves, additional material constants are assumed: the Poisson ratios $v_f = v_m = 0.2$, the differential thermal expansion coefficient $\Delta\alpha = (\alpha_m - \alpha_f) = 3 \times 10^{-6}$ /C; and for $V_f = 0.5$, the matrix thickness between two fibers is set at $t = d$, d being the fiber diameter.

The energy release rate curves are plotted against the possible sizes of the assumed matrix flaw, and for interface flaws of sizes of b = 0, 0.15, 0.25 and 0.5d. It is seen that when b = 0, meaning a perfect interface bonding, the associated energy curve exhibits a maximum value at about a = 0.7d. In this case, the effect of fiber constraint against matrix flaw propagation is most effective. With the presence of the interface flaws, however, the energy release rate at the matrix flaw tip is increased due to loss of fiber constraint. In particular, if the interface flaws are of a size in the order of b = 0.5d, the effect of fiber constraint can be lost completely. If this happens, the threhold stress for matrix flaw propagation will be dramatically decreased.

In the context of the Wang model, then, the energy release rate curves can be used in conjunction with Eq. (4) to determine the critical composite stress at the onset of matrix cracking once the matrix flaw size a and the interface flaw size 2b are specified. The procedure is as follows:

If we assume perfect bonding and set the size of the matrix flaw in the order of 0.7d, then the maximum values of C_e and C_T are taken from the associated curves in Figure 3 and 4. There are valued at 95×10^9 J/m^3 and 0.7 $J/m^3/{}^oC^2$, respectively. Next, from Table 1 we calculate the matrix critical strain energy release rate G_m = 40 J/m^2; and from fabrication data we set ΔT = -700^oC. Then, after applying Eq. (4), the critical composite strain at the propagation of the matrix flaw is calculated as e_c = 3.2×10^{-3}. The composite axial stiffness is E_c = 142 GPa; hence the critical composite stress is σ_c = 456 MPa. This result is to be compared with the experimental value of 375 ± 10 MPa.

Now, if we assume the presence of interface flaws and their size are in the order of the fiber diameter (2b=d), then the corresponding maximum values of C_e and C_T are determined from the corresponding curves in Figures 3 and 4. These are taken as 140×10^9 J/m^3 and 1.1 $J/m^3/{}^oC^2$, respectively. Again, using Eq. (4), the predicted onset strain is 2.3×10^{-3} and the associated onset stress is 328 MPa. This result is also to be compared with the experimental value of 375 ± 10 MPa.

Since the values of b = 0 and 2b = d represent the extremes of the fiber constraint against the propagation of the matrix flaw, the coreesponding onset stresses 456 MPa predicted for b=0 and 328 MPa for 2b = d may be considered as the upper and lower values which bound the matrix cracking stress.

Similarly, the energy coefficient C_e-curves for the SiC/Borosilicate system are shown in Figure 5 (the C_T- curves are omitted due to the small fiber volume fraction in the composite). For this case, the matrix thickness between two fibers is set at t = 4d. The matrix critical energy release rate is G_m= 9 J/m^3. The calculation follows the same procedure as before. Here, we have assumed that the matrix flaw size is in the order of one fiber diameter. By omitting the thermal residual stress effect as being insignificant, the critical matrix cracking stress (upper value) is calculated at 90 MPa. This result is to be compared with the experimental value of 78 ± 10 MPa.

Finally, for the C/borosilicate system the energy coefficients are also computed. But, these are not shown here. Additional material constants used in the calculation are : Poisson ratios v_f = v_m = 0.2, the differential thermal expansion coefficient $\Delta \alpha$ = $3. \times 10^{-6}$ /C and ΔT = 600^oC. The calculated upper and lower values for the matrix cracking stress are 572 MPa and 310 MPa, respectively. This result is to be compared with the experimental value of 360 ± 2 MPa.

CONCLUSION

In this paper, we have presented some predictive models for the matrix cracking stress in brittle ceramic matrix composites. We have also presented some experimental results for the same. Our main purpose here is to illustrate the use of these models, and in particular, the necessary input parameters required for carrying out the calculation. The values of some of the parameters are simply unavailable at the present time, and had to be assumed in order to carry out the calculations. For this reason, no attempt is made to evaluate the adequacy (or the lack thereof) of any of these models, although a brief comparison was made between the predicted and the experimental results. Clearly, more systematic experiments are needed if these models are to be critically assessed.

ACKNOWLEDGMENTS.

The authors would like to thank their students P. V. Kishore and X. Huang for carrying out the computations, and B. Plotnick for conducting the experiment. This work has been supported by a grant from the Air Force Office of Scientific Research.

REFERENCES

1.   Aveston, J., Cooper, G. and Kelly, A, "Single and Multiple Fracture," in The Properties of Fiber Composites, Conference Proceedings, National Physical Laboratory, Guildford, UK. IPC Science and Technology Press Ltd., 1971, pp. 15-26.

2.   Budiansky, B., Hutchinson, J. W. and Evans, A. G., "Matrix Fracture in Fiber-Reinforced Ceramics," J. Mech. Phys. Solids, Vol 34, 1986, pp. 167-189.

3.   Marshall, D., Cox, B. and Evans, A.,"The Mechanics of Matrix Cracking in Brittle-Matrix Fiber Composites," Acta Metall., Vol. 33, 1985, pp. 2013-2021.

4.   McCartney, L. N., "Mechanics of Matrix Cracking in Brittle-Matrix Fiber-reinforced Composites," Proc. Roy. Soc. London, A-409, 1987, pp. 329-350.

5.   Kerans,R.J., Hay, R. S., Pagano, N. J. and Parthasarathy, T. A., "The Role of the Fiber-Matrix Interface in Ceramic Composites," Ceramic Bulletin, Vol. 68, 1989, pp. 429-442.

6.   Kelly, A. and McCartney, L. N., "Matrix Cracking in Fiber-Reinforced and Laminated Composites," Proc. ICCM-6 and ECCM-2, Ed. F. L. Matthews, N. C. R. Buskell, J. M. Hodgkinson and J. Morton, Elsevier Applied Science, London, Vol. 3, 1987, pp. #.210-230.

7.   Wang, A. S. D., "On Fiber-Matrix Interface Bonding and Composite Toughness," Proc. of the 1st USSR-Symposium On Mechanics of Composite Materials, Riga, Latvian SSR, 1989.

8.   Wang, A. S. D.,"Fracture Mechanics of Sublaminate Cracks in Composite Materials," Composite Technology Review, Vol. 6, 1984, pp. 45-62.

9.   Barsoum, M. and Plotnick, B., "Matrix Cracking Stresses in Uniaxially Fiber Reinforced Ceramic Composites," submitted to Jour. Amer. Ceramic Soc.

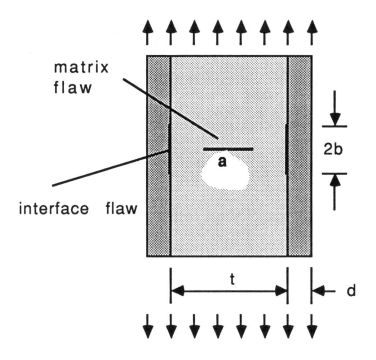

Fig. 1 A unit-cell showing matrix and interface flaws and the fiber-matrix geometry

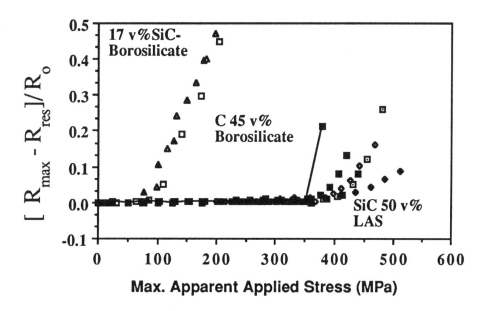

Fig. 2 Measured resistance change plotted against the applied stress

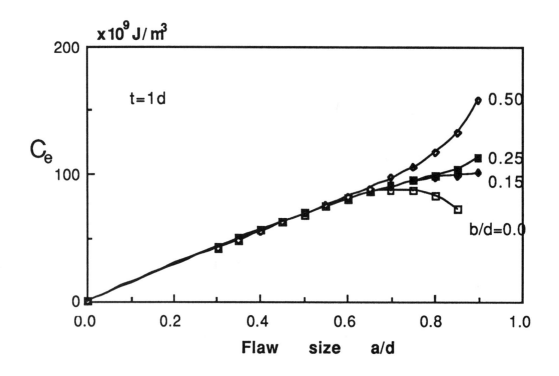

Fig. 3 Strain Energy Release Rate Coefficient, $C_e(a)$, for the SiC/LAS system

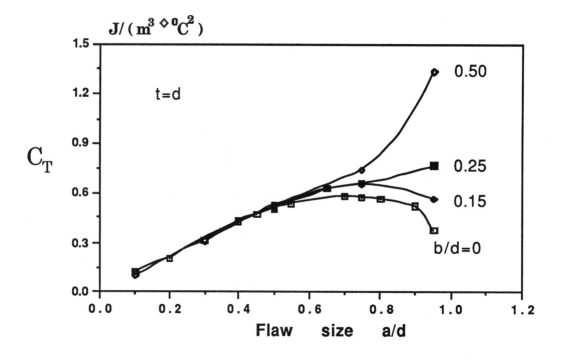

Fig. 4. Strain Energy Release Rate Coefficient, $C_T(a)$, for the SiC/LAS system.

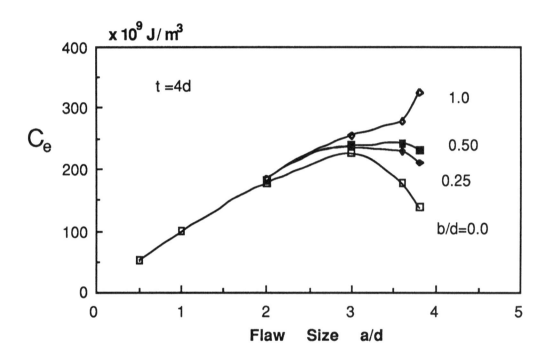

Fig. 5 Strain Energy Release Rate Coefficient, C_e(a) , for the SiC/Borosilicate system.

High Temperature Fatigue of SiC Fiber-Reinforced Si_3N_4 Ceramic Composites

JOHN W. HOLMES,[1] TEMEL KOTIL[1] AND WILLIAM T. FOULDS[2]

ABSTRACT

The tensile fatigue behavior of Si_3N_4, with unidirectional reinforcement by 30 vol% SCS-6 SiC fibers, was investigated at 1000°C. The composite was prepared by conventional hot press techniques. Load-controlled fatigue testing was conducted at a sinusoidal frequency of 10 Hz, and a stress ratio ($\sigma_{min}/\sigma_{max}$) of 0.1. Three maximum stress levels were examined: 180, 220 and 280 MPa.

Under monotonic tensile loading at 1000°C, the composite exhibited a linear stress-strain response up to approximately 200 MPa. Loading beyond 200 MPa produced non-linear behavior, with tensile failure occurring at approximately 380 MPa. For tensile fatigue loading, at a peak stress of 180 MPa, failure was not observed in 2×10^6 cycles. Increasing the maximum fatigue stress to 220 MPa resulted in failure in approximately 270,000 to 749,000 cycles; for a maximum stress of 280 MPa, failure occurred in approximately 4,000 to 33,000 cycles. All fatigue failures were preceded by a gradual decrease in composite modulus. No reduction in modulus was observed for the specimens which survived 2×10^6 cycles at 180 MPa.

Details of an edge-loaded specimen, designed for tensile and fatigue testing of ceramic composites, are given. The 100 mm long specimen provided consistent gage section failures.

INTRODUCTION

Continuously reinforced ceramic composites are under development for use in gas-turbine hot-sections. The successful use of these materials will require extensive characterization of their mechanical behavior and thermophysical properties. Although a sizeable amount of information has been published recently concerning the monotonic fracture behavior and mechanisms of

[1]Assistant Professor and Graduate Student, respectively, Ceramic Composites Research Lab, The University of Michigan, Dept. of Mechanical Engineering and Applied Mechanics, 1065 GGBL, Ann Arbor, MI 48109-2125.
[2]Manager, Ceramic Composite Development, Textron Specialty Materials, Lowell, MA 01851.

toughening for continuously reinforced ceramic composites [1-8], little information is available concerning the elevated temperature fatigue behavior of these materials.   A notable exception is the work of Prewo at United Technologies Research Center. Prewo has published extensively [9-11] on the ambient and elevated temperature tensile- and flexural-fatigue behavior of SiC fiber-reinforced glass-ceramic composites (Nicalon/LAS). In Prewo's work, it was shown that the tensile- and flexural-fatigue life of Nicalon/LAS composites was strongly influenced by the stress at which microcracking of the composite occurs during monotonic loading.   In room temperature tensile-fatigue testing of unidirectionally reinforced Nicalon/LAS-II composites, which exhibit a non-linear monotonic stress-strain curve, progressive fatigue damage was observed for peak stresses above the proportional limit[3] (fatigue damage was inferred by measuring the proportional limit of the composites after 10^5 fatigue cycles, and also by shape changes in the cyclic stress-strain curves). For maximum fatigue stresses below the proportional limit, fatigue damage was not observed within 10^5 cycles. In elevated temperature (600 and 900°C) flexural fatigue testing of unidirectionally reinforced Nicalon/LAS, Prewo [9] noted a gradual decrease in specimen stiffness for fatigue stresses above the proportional limit, indicating the progressive nature of fatigue damage accumulation in continuously reinforced ceramic composites.

The present work provides initial results obtained for the elevated temperature fatigue behavior of a unidirectional SiC-fiber reinforced Si$_3$N$_4$ composite. Particular attention is given to the influence of cyclic stresses above and below the proportional limit on the tensile-fatigue life of the composite.

EXPERIMENTAL PROCEDURE

Material. Si$_3$N$_4$, unidirectionally reinforced with 30 vol.% SCS-6 monofilament SiC fibers, was chosen for this study.[4] The composite was prepared by conventional hot press technology. First, fibers were collimated by applying them to the outside of a revolving drum, where a resin was used to maintain fiber spacing.   The fiber/resin mats were then sectioned from the drum surface and applied sequentially between layers of matrix powder in a graphite mold (dimensions: 108 mm x 108 mm x 8 mm).   The matrix powder formulation was: 93.75% Stark LC-12 Si$_3$N$_4$ (nominal 0.5 μm particle size), 5.00% Y$_2$O$_3$ and 1.25% MgO (all percentages by weight).   The SCS-6 β–SiC fibers had a diameter of 143 μm (Figs. 1a and b). These fibers are manufactured by continuous CVD of β-SiC onto a 33 μm diameter carbon core. The outer diameter of the fiber has a carbon-rich coating, approximately 3 μm thick.   The preform fiber/powder lay up was consolidated at 1700°C under 70 MPa pressure for one hour in a nitrogen atmosphere.   The consolidation was complete, with a measured density of 3.20 g/cm^3 (>99% of the theoretical density).   An SEM micrograph showing typical fiber distribution in the billet is given in Fig. 1b.

---

[3]Defined here as the stress at which initial deviation from linear stress-strain behavior is observed under monotonic loading.
[4]Material supplied by Textron Specialty Materials, Lowell, MA.

Specimen geometry and experimental arrangement.   Tapered-end tensile specimens (Fig. 2) were used for both monotonic and fatigue testing. This edge-loaded specimen was designed with a gage-length of approximately 30 mm. The specimen ends rested in cavities which were machined into attachments for self-aligning grips. Compared to other techniques for specimen gripping which were investigated (e.g., face loading of specimen ends), the edge-loading technique provided the most reproducible specimen alignment. For the self aligning grips and edge-loaded specimen used, specimen bending strains were less than 1% of the axial strain (measured at room temperature in accordance with ASTM standard E1012-84).

Tensile and fatigue testing was conducted on an MTS Model 810 Material Test System. Heating of the specimen gage-length was accomplished using an induction heated SiC susceptor (Fig. 3). The susceptor dimensions were: 52 mm OD x 34 mm ID x 38 mm long (utilizing a thick-walled susceptor ensures that specimen heating, by radiation from the inner-wall of the susceptor, is uniform). When testing ceramic composites in fatigue, where strain amplitudes are extremely low (the order of 0.1%), stable temperature control is crucial to prevent errors in extensometer readings caused by thermal expansion of the specimen and extensometer loading rods. At the test temperature of 1000°C, the control system used in the present investigation maintained specimen temperature to within ±2°C (measured at the *center* of the specimen gage-length). With the center of the specimen at 1000°C, the maximum temperature difference measured *along* the gage-length was 4°C. To prevent heating of the grips by conduction from the specimen, water-cooled Cu heat-sinks were lightly clamped around the specimen ends. In addition to reducing the conductive flow of heat to the grips, the 3 mm thick x 50 mm diameter heat-sinks acted as efficient radiation shields. Utilizing this cooling arrangement, the temperature of the grip attachments (Fig. 3) was reduced to below 150°C (for a specimen temperature of 1000°C).

Currently, commercially available extensometers designed for elevated temperature mechanical testing of ceramics are limited to use at frequencies below 5 Hz. This limitation is due, in part, to the length and low contact-force of the extensometer loading rods.   Thus, to measure fatigue strains at 10 Hz (used in the present investigation), a counter-force extensometer arrangement was designed [12]. This arrangement allowed using a moderate extensometer loading-rod force (150 g per rod); the small bending moment induced in the specimen by the extensometer is cancelled by equal transverse loading on the opposite side of the specimen (Fig. 3).[5] Induction heating of the specimen allowed using relatively short (80 mm) extensometer rods, further improving the frequency response of the extensometer.   The tips of the extensometer and counter-loading rods contacted shallow conical depressions ground into the specimen edge.[6] Extensometer

---

[5]When utilizing this type of arrangement, it is important to verify that the extensometer and counter-force loading rods act along the same centerlines.

[6]The conical depressions served two purposes: (1) to accurately locate the extensometer and counter-loading arms and, (2) to prevent *possible* extensometer slippage due to the energy expended during fiber fracture.   For cross-ply laminates, where edge delamination could occur, the extensometer and counter-loading rods should contact the specimen face, rather than the edges.

loading along the specimen edge was chosen to take advantage of the increased stiffness of the specimen along this direction. The extensometer used in the present study was calibrated to a resolution of ± 0.2 µm. A high-resolution 16-bit data acquisition system was used to monitor the extensometer output.

RESULTS AND DISCUSSION

I. Tensile Tests

To allow establishment of the maximum fatigue loads, the monotonic stress-strain behavior of the SiC/Si$_3$N$_4$ composite was determined at 1000°C. From Fig. 4, the composite exhibited a linear stress-strain response to approximately 200 MPa ($\approx$ 50% of the ultimate strength). Loading beyond 200 MPa produced nonlinear stress-strain behavior, with fracture occurring at approximately 390 MPa. For convenience, the point of *initial* departure from linear stress-strain behavior is defined as the proportional limit stress, σ_{pl}. Physically, σ_{pl} corresponds to the stress at which initial microcracking occurs [13]. Considering data obtained from three tensile tests, the proportional limit stress was 200 ± 18 MPa. The ultimate strength was 390 ± 28 MPa.

II. Fatigue Tests

Load-controlled tensile fatigue testing was conducted at 1000°C, with maximum fatigue stresses that were 90, 110 and 140% of the proportional limit measured at 1000°C. For all stress levels, the stress ratio ($\sigma_{min}/\sigma_{max}$) was maintained at 0.1 and the sinusoidal frequency at 10 Hz. For each stress level, three specimens were tested. The number of cycles to failure versus maximum fatigue stress is plotted in Fig. 5 (see also Table I).

$\sigma_{max} < \sigma_{pl}$. For a maximum fatigue stress of 180 MPa, which is approximately 10% *below* the initial proportional limit of the composite, specimens survived 2×10^6 cycles (defined as run-out in this study). The cyclic stress-strain curve remained linear throughout the testing, with no change in slope. The post-fatigue tensile stress-strain behavior of these specimens was measured at 1000°C to determine if fatigue stresses below the proportional limit influenced the monotonic stress-strain response of the composite. From data obtained on three specimens, the proportional limit and ultimate strength ranged from 190 to 212 MPa and 368 to 395 MPa, respectively (Table 1). From the results obtained on this limited number of specimens, it appears that composite strength and modulus are not affected by fatigue cycling below the initial proportional limit of the composite.

$\sigma_{max} > \sigma_{pl}$. For maximum fatigue stresses above the proportional limit, fatigue failures were observed . For a maximum fatigue stress of 220 MPa, failure occurred in approximately 270,000 to 749,000 cycles; at 280 MPa, failure was observed in approximately 4,000 to 33,000 cycles. As shown in Figs. 6a and b, failure was preceded by a gradual decrease in specimen stiffness. This decrease in stiffness was accompanied by considerable hysteresis in the cyclic stress-strain curves.   In addition to an increase in strain amplitude caused by the increase in specimen compliance, considerable ratchetting of the cyclic stress-strain curves was observed (ratchetting was not observed for fatigue with $\sigma_{max} < \sigma_{pl}$).   This ratchetting can be attributed to extension of the specimen gage section due to fiber/matrix cracking and pullout.   A contribution from creep to strain ratchetting does not appear likely (at 1000°C), since a similar shift was also observed for specimens which had failed at 270,430 and 749,326 cycles (approximately 7.5 and. 21 hr at 1000°C, respectively).   As shown in Fig. 7, only a moderate amount of fiber pullout accompanied the fatigue failures. The degree of fiber pullout was similar to that obtained with monotonic tensile loading.

CONCLUSIONS

1. The tensile-fatigue behavior of hot pressed Si_3N_4, with unidirectional reinforcement by monofilament SiC fibers, was investigated at 1000°C under a sinusoidal frequency of 10 Hz.   For a maximum fatigue stress of 180 MPa (approximately 10% below the initial elastic limit of the composite) failure was not observed in 2×10^6 cycles.   Measurement of composite strength and modulus after fatigue testing showed that no discernable change in these properties had occurred as a result of fatigue loading.   These results are encouraging, since many potential applications of fiber-reinforced ceramic composites will be in the high cycle fatigue regime.

2. For a maximum fatigue stress of 220 MPa (approximately 10% above the initial elastic limit of the composite), failure occurred in approximately 270,000 to 749,000 cycles. At 280 MPa, fatigue failures were observed in approximately 4,000 to 33,000 cycles.   For maximum stresses of 220 and 280 MPa, a gradual decrease in composite modulus and pronounced hysteresis in the cyclic stress-strain curves was observed, indicating the progressive accumulation of fatigue damage in the composite.   These gradual fatigue failures, observed for stresses above the initial elastic limit of the composite, indicate that it *may* be possible to use continuously reinforced ceramic composites in non-critical applications involving cyclic loading above the initial elastic limit of the composite.

3. A 100 mm long tensile specimen, with a gage length of approximately 30 mm, has been developed for mechanical testing of ceramic composites.   The edge-loaded specimen showed consistent gage section failures for both monotonic and fatigue loading.

4. The results presented in this paper are based on a limited number of tests (3 specimens per loading condition); further work is required to determine the statistical variation in results which would be observed with a larger population of test specimens.   Additional work is also needed to determine the influence of time-at-temperature and accompanying interfacial reactions on fatigue life.

REFERENCES

1. Luh, E. Y. and Evans, A. G., "High-Temperature Mechanical Properties of a Ceramic Matrix Composite," *J. Am. Ceram. Soc.,* **70** [7] (1987), 466-469.

2. Moeller, H. H., Long, W. G., Caputo, A. J. and Lowden, R. A., "Fiber Reinforced Ceramic Composites," Advanced Materials Technology '87, (1987), 1519-1527 (Publ.: Society for the Advancement of Material and Process Engineering, Covina, CA 91722).

3. Hillig, W. B., "Strength and Toughness of Ceramic Matrix Composites," *Annual Review of Materials Science,* Vol. 17, (1987), 341-383.

4. Thouless, M. D. and Evans, A. G., "Effects of Pull-Out on the Mechanical Properties of Ceramic-Matrix Composites," *Acta. Metall.,* **36** [3] (1988), 517-522.

5. Marshal, D. B. and Cox, B. N., "Tensile Behavior of Brittle Matrix Composites: Influence of Fiber Strength," *Acta Metall.,* **35** [11] (1987), 2607-2619.

6. Hopkins, G. R. and Chin, J., "SiC Matrix/SiC Composite: a High-Heat Flux, Low Activation Structural Material," *J. Nucl. Mater,* Nov.-Dec. 1986, 148-151.

7. Sutco, M., "Statistical Fibre Failure and Single Crack Behavior in Uniaxially Reinforced Ceramic Composites," *J. Material. Sci.,* **23** (1988), 928-933.

8. Pagano, N. J. and Dharani, L. R., "Failure Modes in Unidirectional Brittle Matrix Composites (BMC)," *Ceram. Eng. Sci. Proc.,* **8** [7-8] (1987), 626-629.

9. Prewo, K. M., "Fatigue and Stress Rupture of Silicon Carbide Fibre-Reinforced Glass-Ceramics," *J. Mater. Sci.,* **22** (1987), 2695-2701.

10. Nardone, V. C. and Prewo, K. M., "Tensile Performance of Carbon-Fibre-Reinforced Glass," *J. Mater. Sci.,* **23** [1] (1988), 168-180.

11. Minford, E. and Prewo, K. M. "Fatigue of Silicon Carbide Reinforced Lithium-Alumino-Silicate Glass-Ceramics," in Tailoring Multiphase and Composite Ceramics, C. G. Patano and R. E. Messing, Eds., Plenum Publishing Corporation (1986), 561-570.

12. Kotil, T. and Holmes, J. W., "A Technique for Measuring High-Frequency Elevated Temperature Fatigue Strains in Ceramic Composites," in preparation.

13. Sakamoto, H., Hironori, H. and Miyoshi, T., "In Situ Observation of Fracture Behavior of SiC Fiber-Si$_3$N$_4$ Matrix Composite," *J. Ceram. Soc. Jpn. Inter. Ed.,* **95**, (1987), 817-822.

Table 1. Maximum fatigue stress, cycles to failure and residual strength for fatigue testing of Si$_3$N$_4$/SiC composites. All experiments were conducted at 1000°C, with a stress ratio of 0.1 and sinusoidal frequency of 10 Hz. The proportional limit of the composite σ_{pl} at 1000°C was approximately 200 MPa; the ultimate strength σ_{ult} was approximately 380 MPa.

Max. Fatigue Stress, σ_{max}	σ_{max}/σ_{pl}	$\sigma_{max}/\sigma_{ult}$	Cycles to Failure	Residual Strength
280 MPa	1.4	0.74	4,270	---
"	"	"	10,700	---
"	"	"	33,321	---
220	1.1	0.58	270,430	---
"	"	"	379,410	---
"	"	"	749,326	---
180	0.9	0.47	$>2 \times 10^{6*}$	368 MPa
"	"	"	$>2 \times 10^6$	382
"	"	"	$>2 \times 10^6$	395

*Defined as run-out for this initial investigation.

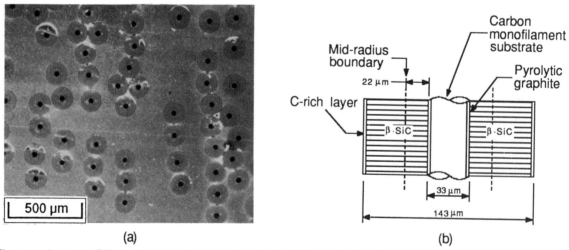

(a)   (b)

Figs. 1a,b   (a) SEM micrograph showing typical fiber distribution in the hot pressed SiC-fiber reinforced Si$_3$N$_4$ composite which was studied.   (b) Schematic of SCS-6 SiC fiber showing approximate size of carbon core and CVD SiC.

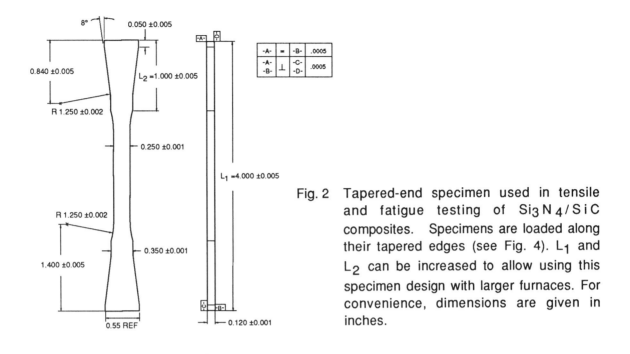

Fig. 2 Tapered-end specimen used in tensile and fatigue testing of Si_3N_4/SiC composites. Specimens are loaded along their tapered edges (see Fig. 4). L_1 and L_2 can be increased to allow using this specimen design with larger furnaces. For convenience, dimensions are given in inches.

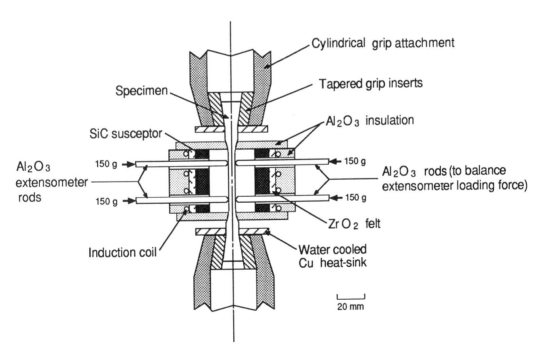

Fig. 3 Schematic of induction heated furnace and specimen gripping arrangement used for tensile and fatigue testing of ceramic composites. A 2.5 kW (450 kHz) induction generator was used to heat the SiC susceptor. Cu heat-sinks were used to reduce heat flow to the grips. The conical-point extensometer loading arms were 5 mm in diameter x 80 mm long.

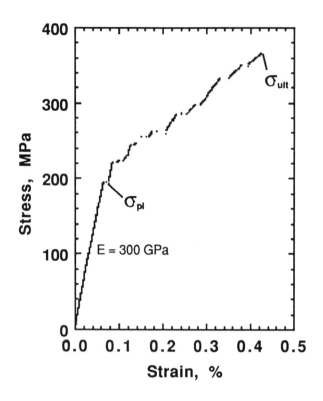

Fig. 4   Typical monotonic stress-strain curve for hot pressed SiC fiber-reinforced Si_3N_4 at 1000°C.

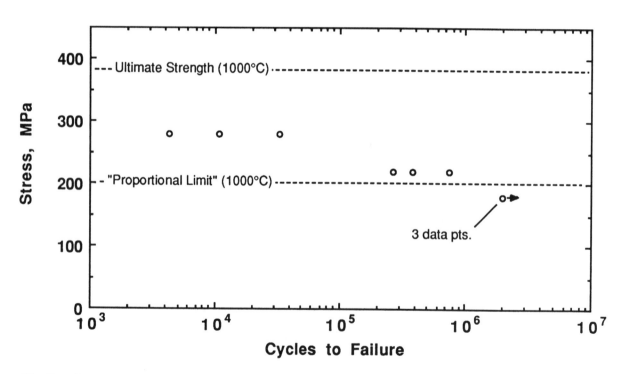

Fig. 5   Cycles to failure versus maximum fatigue stress for SiC fiber-reinforced Si_3N_4. The average proportional limit and ultimate strength at 1000°C are shown for comparison.

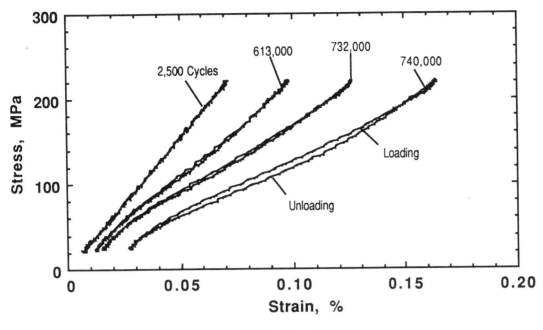

(a) σ_{max} = 220 MPa, N_f = 749,326 cycles.

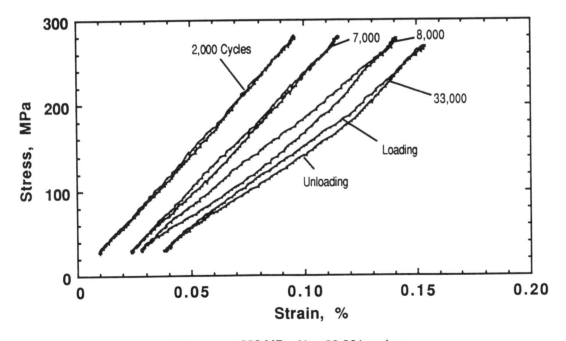

(b) σ_{max} = 280 MPa, N_f = 33,321 cycles.

Figs. 6a,b. Cyclic stress-strain behavior of SiC fiber-reinforced Si_3N_4 during fatigue loading at 1000°C. Ratchetting of the stress-strain curves is attributed to fiber/matrix cracking and pullout, which produced extension of the specimen gage section (in addition to extensometer data, this extension was verified by monitoring output of the displacement transducer).

200 µm

Fig. 7   SEM micrograph showing fracture
surface obtained with fatigue loading of
SiC fiber-reinforced Si_3N_4 composite
(σ_{max} = 280 MPa, cycles to failure =
33,321).

A Micro-Laminate Effect in Brittle Matrix Composites

N. J. PAGANO   **R. Y. KIM**

ABSTRACT

The influence of fiber-matrix interfacial debonding upon laminate behavior is examined through an analytical/experimental study for a glass matrix composite (Nicalon-1723 glass). The NDSANDS model in conjunction with classical lamination theory is used to determine the micro- and macro-stresses in ± 45° laminates subjected to residual curing stresses followed by tensile or compression loadings. Experiments are then conducted to examine the modeling assumptions. Of particular concern are the onset of debonding and subsequent response. It is postulated that debonding is caused by the relief of the initial radial compressive stress at the fiber-matrix interface. The results demonstrate the potential impact of micromechanical details on the response of brittle matrix composite laminate.

INTRODUCTION

In this work we shall consider the behavior of a laminated brittle matrix composite material (BMC) under the action of both residual curing stresses and applied loading conditions. Our emphasis is on the influence of fiber-matrix debonding* upon the elastic and failure response of a laminated system. In particular, recent experimental results [1] have shown that the composite transverse modulus may be extremely sensitive to fiber-matrix interface conditions. In that study, the transverse moduli of two composites having the same fiber (Nicalon) but different matrices (CAS and LAS) were measured. The two

*The term "debonding" as used here simply means a loss in contact has taken place at the interface between two materials. It does not imply that a physical failure process has occurred, such as the breaking of chemical bonds. In fact, the case considered here represents (theoretical) interface separation due to relief of an initial radial compressive stress. We also assume in this work that the composite can be modeled by considering only two materials, fiber and matrix, as we have no information regarding mechanical properties of potential interphase regions.

N. J. Pagano, Materials Research Engineer, WRDC/MLBM, Wright-Patterson AFB OH 45433-6533
R. Y. Kim, University of Dayton Research Institute, Dayton OH

matrix materials have approximately equal Young's module but CAS has a coefficient of expansion greater than that of the fiber, while for LAS the reverse is true. Therefore, modeling the residual stress development as a thermoelastic cooldown problem leads to a compressive radial stress at the interface in CAS-Nicalon and, assuming perfect bond, tensile radial stress in LAS-Nicalon. Hence the latter is susceptible to interfacial debonding and the attendant lower composite transverse modulus. Experiments confirmed this behavior with the CAS- and LAS-Nicalon transverse module being reported as 110 GPa and 21 GPa, respectively. In turn, the measured properties were in close agreement with theoretical predictions given by NDSANDS[2], where the theoretical value for the LAS-Nicalon is based upon the upper bound calculation for a composite with voids replacing the fibers.

The foregoing discussion leads one to inquire regarding the influence of micromechanical phenomena, such as debonding, on the behavior of a laminated system. In such a body the (macroscopic) stresses acting on the layers are dependent upon the layer properties, orientations, and stacking sequence, as well as the applied loading. By "macroscopic" stresses we mean the volumed averaged stresses within the constituents of a given layer. In turn, the fiber, matrix, and interface response depend upon the macroscopic stresses.

Thus we cannot claim a'priori knowledge of the layer moduli as one does for continuous interface composites. Rather, an iterative procedure may be required to execute layer-by-layer laminate stress field analysis. We shall refer to this interdependence of properties as a micro-laminate effect. The final predicted stress state within a laminate must be consistent with the physical behavior of the interface and the details of the laminate loading and geometry. In this work we shall examine a laminate which demonstrates this type of behavior.

ANALYTICAL/EXPERIMENTAL STUDY

Our objective is to demonstrate the dependence of laminate response on interfacial conditions (the micro-laminate effect) and to examine the capabilities of a contemporary analytical model [2], in conjunction with classical lamination theory [3,4,5], to quantify this effect. Specifically, $\pm 45°$ laminates of Nicalon-1723 glass undirectional layers shall be submitted to experimental/analytical investigation.

To demonstrate the micro-laminate effect in a realistic system, we have conducted experiments on $\pm 45°$ symmetrics laminates under both tensile and compressive loadings.

The fiber volume fraction is 0.35. Load and strain data is converted to τ_{LT} vs. γ_{LT}, the macroscopic shear stress and engineering shear strain, respectively, of the individual layers in the L (fiber) and T (transverse) directions. This is done since the $\pm 45°$ laminate may receive consideration as a shear test for BMC and also because the analytical model utilized [2,6] leads to coincident upper and lower founds for the in-plane shear modulus, G_{LT}, of a layer. We must note, however, that this property is expected to be strongly influenced by interfacial characteristics, and should, therefore, be viewed as a function of the imposed stress/strain, rather than as a constant.

We have not characterized the individual constituents of the composite. Hence we appeal to data cited by Grande et al [7] as input for our micromechanical computations. These nominal properties are

	Nicalon	1723 Glass
E(GPa)	200	88
G(GPa)	77	36
α (μm/m°C)	3.2	5.2

where E and G are room temperature moduli and each constituent is assumed to be isotropic. The values reported for α correspond to averages over the range from room temperature to the glass solidification temperature (approx. 600°C).

We begin by computing the effective layer thermoelastic properties corresponding to a continuous fiber-matrix interface by use of NDSANDS in conjunction with the nominal properties above. By application of classical lamination theory then, we calculate the stress and strain components (away from the free edge region) within each layer for conditions consisting of

a) A 500°C drop in temperature of the entire laminate to simulate the development of curing stresses, and

b) A unit applied axial normal stress on the laminate.

For each condition, a) and b), we return to NDSANDS to determine the stress fields within the constituent materials. The entire procedure is repeated for complete interfacial separation at the fiber-matrix interface, since, even though the curing leads to compressive radial stress, we expect this to be relieved in the tension loading of the laminate. The temperature drop of 500°C was arbitrarily chosen to represent the cooldown phase of processing.

The computed curing stress distributions of the radial, hoop, and axial normal stresses in the matrix material at the fiber-matrix interface are shown in figs 1, 2, and 3,

respectively. The shear stress $\tau_{r\theta}$ is also non-zero, but has a fairly small magnitude. The radial stress distribution due to a tensile loading of 1MPa is shown in fig. 4. The remaining 5 stress components are non-zero but are not shown here. According to figs. 1 and 4, the radial stress of -34.5 MPa would be relieved at an applied stress level of 59.5 MPa. While the interface debonding would be expected to grow gradually, our present

models cannot account for this level of sophistication. Consequently, the theoretical τ vs. γ plot is displayed in fig. 5 as a bilinear curve with the knee at 29.8 MPa (the magnitude of

τ_{LT} due to loading is one-half of the applied laminate stress). The experimental result is also shown for comparison in fig . 5. It should be noted that the previous prediction assumes that debonding is the only failure mode within the composite. The prediction of ultimate failure is not attempted. The high tensile values of hoop and axial normal stress shown in figs. 2 and 3 may also lead to failure(s) as the bulk matrix strength is on the order of 60 MPa. Actually, numerous initial cracks were observed both parallel and perpendicular to the fiber direction (fig. 6) however, subsequent machining demonstrated that this damage only occurred in a glass-rich region at the surface of the specimen. Other damage in the form of transverse cracks on the free edges (fig. 7) may be responsible for the continuously degraded tangent modulus of the actual stress-strain curve. The ultimate failure mode, which involves interlaminar as well as in-plane damage, is shown in fig. 8.

Evidently, the postulated interfacial debonding leads to a significant impact on the laminate response in the previous example. Further evidence of this micro-laminate effect can be seen by comparison of the results for tensile and compressive loading of the ± 45° laminate. This comparison is shown in fig. 9, in which the tensile loading curve has been drawn to a different scale from that of fig. 5. In this figure we can observe a much greater retention of elastic stiffness (tangent modulus) and a delayed "proportional limit" in the compression sample. This is consistent with the postulated mechanism of non-linearity since no debonding (at least that caused by radial <u>normal</u> stress) is predicted for the compression loading. Unlike the tension case, no transverse cracks were observed in compression. Hence the mechanism responsible for the non-linearity in compression has not been identified at this time.

SUMMARY

We have taken an initial look at the credibility of contemporary analytical models to study the influence of micromechanical parameters, particularly interfacial debonding, on laminate response. Indirect evidence supports the notion of a "micro-laminate effect" indicating that ply or layer elastic properties may not be treated as unique quantities, but are strong functions of interface behavior, which, in turn, is dependent on macroscopic state of stress. Tension and compression tests on ± 45° laminates have displayed a consistent picture with regard to this phenomenon. The apparent degree of predictability may be surprising in view of the internal variability one sees in the constituent material geometry of glass and glass-ceramic matrix composites, but leaves one with an optimistic projection for subsequent modeling successes. While the present theoretical/experimental correlation has been promising, much more research is needed to provide a more descriptive model of the debonding process, as well as in the study of other layer and laminate failure modes.

ACKNOWLEDGEMENT

The authors wish to express their gratitude to Dr G. P. Tandon of AdTech Systems Research Inc for providing the analytical results that were needed for this study.

REFERENCES

1. R. Y. Kim and N. J. Pagano "Initiation of Damage in Undirectional Brittle Matrix Composites" Proceedings of the 1988 U.S./Japan Conference on Composite Materials, American Society for Composites, Technomic, USA (1988).

2. N. J. Pagano and G. P. Tandon "Elastic Response of Multidirectional Coated-Fiber Composites," <u>Composites Science and Technology</u>, Vol 31 (1988).

3. E. Reissner and Y. Stavsky "Bending and Stretching of Certain Types of Heterogeneous Aeolotropic Elastic Plates" <u>J. Applied Mechanics</u> Vol 28 (1961).

4. S. B. Dong, K. S. Pister, and R. L. Taylor "On the Theory of Laminated Anisotropic Shells and Plates" <u>J. Aerospace Sciences</u> Vol. 29 (1962).

5. J. M. Whitney "Structural Analysis of Laminated Anisotropic Plates," Technomic, USA 1987).

6. N. J. Pagano and G. P. Tandon, to be published (1989).

7. D. H. Grande, J. F. Mandell, and K. C. C. Hong "Fiber/Matrix Bond Strength Studies of Glass, Ceramic, and Metal Matrix Composites," <u>J. Materials Science</u> (1987).

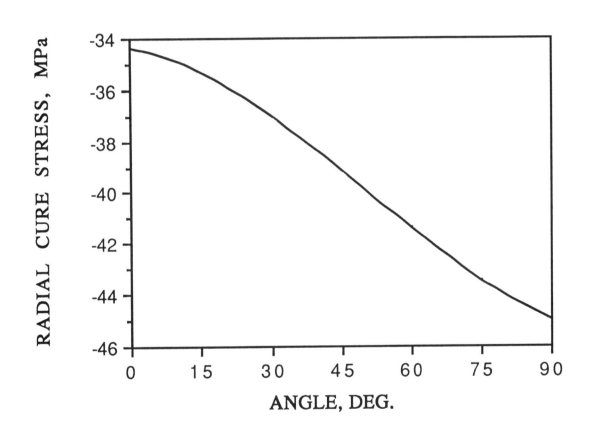

Fig. 1. Radial Cure Stress at Fiber-Matrix
Interface in ± 45° Laminate

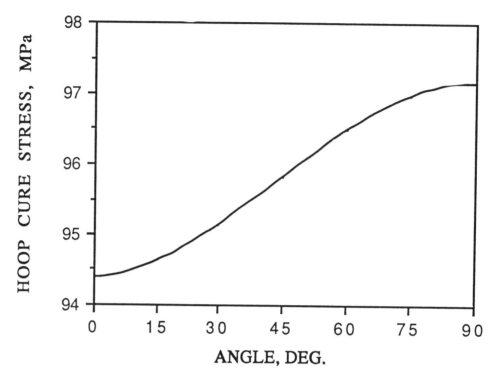

Fig. 2. Hoop Cure Stress at Fiber-Matrix
Interface in Matrix of ± 45° Laminate

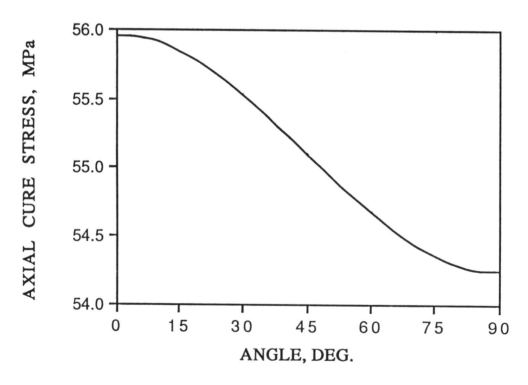

Fig. 3. Axial Cure Stress at Fiber-Matrix
Interface in Matrix of ± 45° Laminate

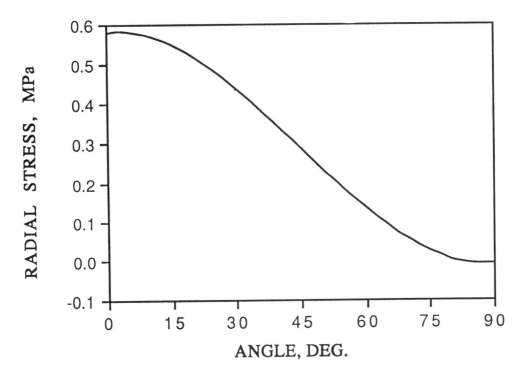

Fig. 4.   Radial Stress at Fiber–Matrix Interface Due to
Tensile Loading of 1 MPa on ± 45° Laminate

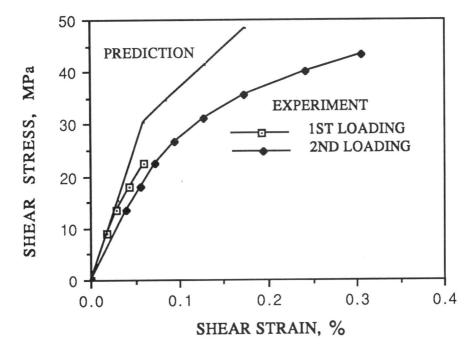

Fig. 5.   Analytical vs. Experimental Shear Stress–Strain
Curve for Tension Loading of ± 45° Laminate

100μm

FREE EDGE

FLAT SURFACE

Fig. 6. Photomicrographs of Initial Matrix Cracks

194

LOADING
←→

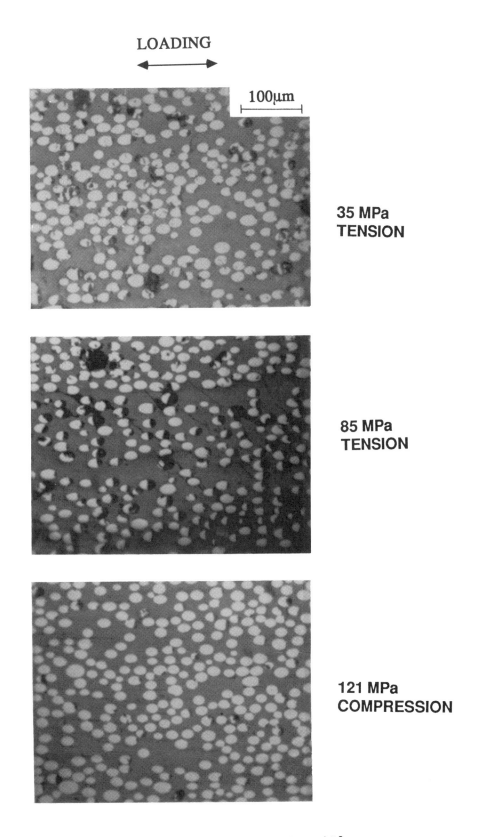

100μm

35 MPa
TENSION

85 MPa
TENSION

121 MPa
COMPRESSION

Fig. 7. Photomicrographs of Edge Surface of ± 45°
Laminates Under Tension and Compression

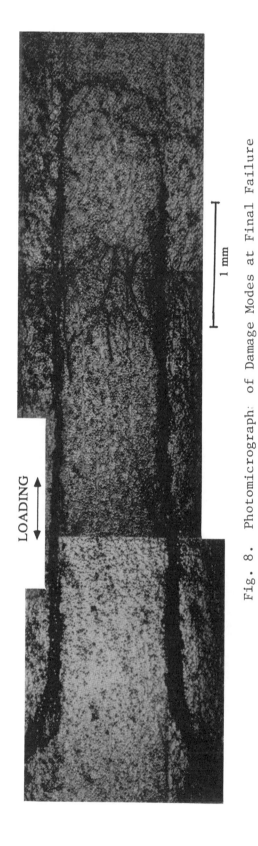

LOADING

1 mm

Fig. 8. Photomicrograph of Damage Modes at Final Failure

196

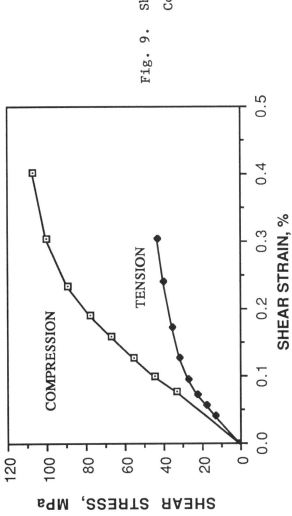

Fig. 9. Shear Stress-Strain Curves for Tensile and
Compressive Loadings of $\pm$ 45° Laminate

197

Anomalous Expansion Behavior in Ceramic Fiber Reinforced Brittle Matrix Composites

A. CHATTERJEE,* G. P. TANDON* AND L. E. MATSON†

ABSTRACT

Analytical modeling of the coefficients of thermal expansion of unidirectional fiber reinforced composite materials shows that the transverse coefficient of thermal expansion of the composite can be higher than that of its constituents at low fiber volume fractions. This effect is especially noticeable if the composite is fabricated with fibers of high modulus and low thermal expansion coefficient in matrices of low modulus and high thermal expansion coefficient. An experimental investigation was therefore carried out to study this anomalous behavior in a Textron fiber (SCS-6) reinforced Hercules 3501-6 epoxy matrix. Numerical calculations for this system have shown that increases of the order of 20 % over the matrix expansion coefficient is possible for fiber volume fractions in the range of 4-5 %. An experimental setup was therefore designed to make perfectly aligned unidirectional specimens using the silicon carbide monofilaments at low fiber volume fractions. The transverse coefficient of thermal expansion was measured using a thermo-mechanical analyzer. Preliminary observations seem to be in favourable agreement with the theoretical calculations.

INTRODUCTION

The design of a fibrous composite material to yield the properties desired for a particular application requires analytical relations between the effective properties of the composite and mechanical and geometric properties of its constituents. Consideration of thermoelastic problems for composites requires the definition of composite moduli, thermal conductivities, specific heat and thermal expansion coefficients. A comprehensive review of composite thermoelastic properties is included in Hashin [1].

In the present paper attention is focussed on the effective thermal expansion coefficients of a two-phase unidirectional fibrous composite. This is defined as a material containing a parallel set of fibers embedded in a homogeneous matrix material. Evaluation of the response of composite materials to temperature changes is important not only for high and low temperature applications but also for fabrication considerations such as the choice of cure temperature for fiber reinforced composites. Thermal expansion behavior can also be important when composite materials are used in conjunction with other materials and when it is necessary to

---

* AdTech Systems Research Inc., Dayton , OH.
† WRDC / MLLM, WPAFB, OH.

match the thermal expansion coefficient of one structural component with another for dimensional stability and mechanical compatibility. From an engineering standpoint, the understanding of thermal expansion coefficients of unidirectional composites has become significant because of the wide use of composites in various applications in recent years. Thermal residual stresses in laminates cannot be calculated without full information about the thermal expansion behavior and elastic response of unidirectional composites.

The effective thermal expansion coefficients are the average strains resulting from a unit temperature rise for a traction free material. Effective thermal expansion coefficients of anisotropic composites having any number of anistropic phases have been bounded from above and below using thermoelastic energy principles by Rosen and Hashin [2]. For isotropic phases, the thermal expansion results reduce to the bounds obtained by Schapery [3]. When the composite has only two phases, thermal expansion coefficients coincide to give exact solutions. The two phase results for thermal expansion coefficients can be obtained more directly following the methods of Levin [4]. For the unidirectional fibrous composite (1- fiber direction) of two isotropic phases, there are two different expansion coefficients (the axial, α_{11} , and the transverse, α_{22}) given by

$$\alpha_{11} = \{ c_f \alpha_f + c_m \alpha_m \} + \frac{\alpha_f - \alpha_m}{\frac{1}{K_f} - \frac{1}{K_m}} \left[\frac{3 (1 - 2 v_{12})}{E_{11}} - \left\{ \frac{c_f}{K_f} + \frac{c_m}{K_m} \right\} \right]$$

$$\alpha_{22} = \{ c_f \alpha_f + c_m \alpha_m \} + \frac{\alpha_f - \alpha_m}{\frac{1}{K_f} - \frac{1}{K_m}} \left[\frac{3}{2 K_{23}} - \frac{3 v_{12} (1 - 2 v_{12})}{E_{11}} - \left\{ \frac{c_f}{K_f} + \frac{c_m}{K_m} \right\} \right]$$

where E_{11}, K_{23}, and v_{12} are the composite effective uniaxial modulus, effective plain strain bulk modulus, and effective axial Poisson's ratio, respectively; c , the phase volume fraction, α, the coefficient of thermal expansion , K, the bulk modulus and the subscripts f and m refer to the fiber and matrix, respectively.

Interestingly, the two phase effective coefficients of thermal expansion do not require the introduction of a specific geometric model. They depend only on the symmetry characteristics of the equivalent homogeneous medium. However, they do have a dependence on the effective elastic properties, which must be known either from experimental data or from the theoretical prediction based on a particular geometric model. Thus, when the solutions are available for the elastic moduli, the problem is solved for the two-phase composite.

Theoretical predictions of the effective coefficient of thermal expansion by Rosen [5] using the composite cylinder model [6] have indicated that the transverse coefficient of thermal expansion of the composite can be higher than that of its constituents at low fiber volume fractions. Similar observations have been made by Chamis and Sendeckyj [7] and Schapery [3] amongst other researchers. This effect is especially noticeable with fibers of high modulus and low axial expansion coefficient (e.g., boron or carbon) in a low-modulus matrix having a high coefficient of thermal expansion (e.g., epoxy resin) [8]. The coefficients of thermal expansion have also been computed by Ishikawa et al [9] based on a hexagonal and square array model and have reported observing a maximum in the transverse expansion coefficient at a very low fiber volume fraction for both carbon and glass fiber reinforced epoxy composites. Using an infinite series solution based upon the known local elastic field solution derived by the perturbation expansion of the Green's tensor function, Nomura and Chou [10] have derived the effective thermal expansion coefficients for multi-phase short fiber composites and have numerically shown that a maximum of transverse expansion coefficient occurs at low fiber volume fraction for continuous glass fiber reinforced epoxy composites. The effective thermal expansion coefficients of an aligned short fiber composite have also been investigated by Takao and Taya [11,12] using Eshelby's equivalent inclusion method and have shown that a peak occurs at fiber volume fraction approximately equal to 0.1 in the carbon/epoxy system for large fiber aspect ratios (i.e., continuous fibers).

Some limited experimental measurement of the transverse coefficient of thermal expansion has been done by Schneider [13] and Yates et al [14], amongst others, at high fiber volume fractions (~ 0.3 - 0.8). To the writers knowledge, the experimental data for the thermal expansion coefficient at low fiber volume fractions is lacking.

The purpose of this work was therefore to study experimentally the transverse thermal expansion behavior of unidirectional fiber reinforced composites at low fiber volume fractions (< 0.1) and to test the applicability of current methods of analysis. We were specifically interested in examining the peak in the behavior of the transverse coefficient of thermal expansion at low fiber volume fractions. The material system that we chose for this study consisted of high modulus Textron fiber (SCS-6) reinforced Hercules 3501-6 epoxy matrix. Unidirectional specimens consisting of controlled fiber volume fractions were made and the effective thermal property measured.

EXPERIMENTAL PROCEDURE

The composites were fabricated using a tetraglycidyl diamino diphenyl methane cured with diamino diphenyl sulfone (H-3501-6) resin manufactured by Hercules Corp. and CVD silicon carbide fibers SCS-6 manufactured by Textron Specialty Materials Corp. Selected properties of the fiber and the matrix are shown in Table 1. The choice of using monofilaments rather than yarns was based primarily on the ability to control the spacings and the volume fraction of the composites in a predictable fashion, have uniformity in fiber dimensions as well as having perfectly aligned unidirectional composites with low volume fractions of the fiber. Specially designed aluminum jigs were machined with closely spaced holes shown in Fig.1. Several of these jigs were made so as to be able to vary the volume fraction of the fibers from 1 - 8 %. The fibers were cut to 50 mm lengths and cleaned in acetone thoroughly to remove all the sizing. The fibers were then inserted in the jig depicted in Fig. 2. The loaded jig was then put in a silicone rubber mold which had already been outgassed so as to eliminate porosity in the resulting composites.

The resin was liquefied at 120 ^{0}C and vacuum debulked for 30 minutes. The molten resin was transfered into the mold containing the fibers and cured at 120 ^{0}C for another 30 minutes. Then the composite was cured at 180 ^{0}C for 2 hours and then slowly cooled to room temperature. A typical cooling cycle would take several hours. This was done to avoid thermal shock and subsequent matrix cracking. Excess matrix was pored into the mold so as to have specimens containing only the matrix made from the same batch of the resin as the actual composites. For every pouring the thermal expansion coefficient of the matrix was measured along with that of the corresponding composites. This was done to ensure that the comparisons between the matrix and the composite properties were made on specimens cured in an identical fashion.

The sample was then taken out of the rubber mold and the aluminum jig was cut off using a high speed water cooled diamond saw. The excess matrix sticking on to the jig was burnt out at 375 ^{0}C for 45 minutes and could be reused again. Then specimens measuring 12 x 4 x 5 mm were cut from this composite block for the measurement of the transverse thermal properties. To get the highest sensitivity from the measurements the specimens had its longest dimension transverse to the fiber direction shown in Fig. 3. The specimens containing just the matrix had the same dimensions as the composites. All specimens were cut using a slow speed diamond saw and polished succesively upto 600 grit to have flat parallel faces and also to remove any other defects.

The transverse thermal expansion coefficients were measured using a Du Pont Thermo Mechanical Analyzer Model 9900 interfaced to a computer for data collection and analysis. The machine was calibrated using a Tungsten NBS standard before use. From the softening point data on the matrix all thermal expansion measurements were made to temperatures of 175 ^{0}C. All tests were done in air. The heating cycle was programmed so as to have soaks at regular intervals of 25 ^{0}C upto the maximum temperature. This was done to ensure that the sample had sufficent time at the temperature of interest to equilibrate. The heating rate was chosen to be 2 ^{0}C /min. The thermal expansion coefficients were calculated by the computer using a point to point line instead of a best fit between the selected temperaures. Since the composites were transparent, the volume fraction of the fibers were measured by counting the number of monofilaments in each specimen and multiplying it with the cross sectional area of each fiber and dividing that by the cross sectional area of the composite.

RESULTS AND DISCUSSION

With the fibers aligned undirectionally (1 - the fiber direction), the composite is transversely isotropic and has two independent thermal expansion coefficients, namely, in the axial and transverse directions. The composite material system that we have used for analysis consists of SCS-6 fibers of high elastic modulus and low thermal expansion coefficient in a H 3501-6 epoxy matrix of low elastic modulus and high thermal expansion coefficient. The concentric cylinder assemblage model developed by Pagano and Tandon [15] has been used to calculate the effective thermal expansion coefficients. The theoretical results for the transverse thermal expansion coefficient as a function of the fiber volume fraction have been plotted in Fig. 4. and show that at very low fiber volume fractions the transverse coefficient of thermal expansion of the composite initially increases beyond that of the matrix value. As seen from the figure, the transverse thermal expansion coefficient shows a peak at a fiber volume fraction of 0.035 approximately. However, at larger fiber volume fractions, this trend reverses and the transverse coefficient of thermal expansion decreases with increasing volume fraction of the fiber. The experimental data has also been plotted on the same figure (The theoretical curve has been shown as a solid line and the data points are the experimental values). The samples which have been fabricated so far are at two volume fractions namely 0.02 and 0.04. As evident from the plot there seems to be good correlation between the data and the theoretical predictions at the fiber volume fraction of 0.02. At the higher volume fraction of 0.04, the experimentally measured transverse thermal expansion coefficient is slightly higher than the numerical prediction. This is, however, within limits of experimental error and is probably because of the variability in the matrix properties from batch to batch.

Nevertheless, the experimental measurements of α_{22} are certainly greater than that of the matrix at the two fiber volume fractions investigated. Experiments are also under way to fabricate specimens at intermediate and higher fiber volume fractions (5-25%) to examine the decreasing trend within this range.

As pointed out by Schapery [3], the initial increase in the transverse coefficient of thermal expansion is apparently due to the axial restraint of the fibers since the matrix is practically in a state of plane strain, except at very small fiber volume fractions. In the longitudinal direction the composite properties are predominately fiber dominated. Since the fiber has a low coefficient of thermal expansion and a high elastic modulus, the matrix in the composite is constrained from expanding in the longitudinal direction even though it (matrix) has a high coefficient of thermal expansion. However, in the transverse direction the matrix is free to expand. Therefore the longitudinal constraint felt by the matrix is compensated by larger expansion in the transverse direction. With increasing volume fraction of fibers, the fiber properties begin to eventually dictate the expansion behavior of the composite. Since the fiber has a low coefficient of thermal expansion, the overall composite property decreases. Further investigation to explain this paradox using a laminate analogy is currently being undertaken and will be reported elsewhere.

SUMMARY

Theoretical predictions of the coefficients of thermal expansion of unidirectional fiber reinforced composite materials shows that the transverse coefficient of thermal expansion of the composite can be higher than that of its constituents at low fiber volume fractions. An experimental investigation was therefore carried out to study this anomalous behavior in Textron fiber (SCS-6) reinforced Hercules 3501-6 epoxy matrix. The transverse coefficient of thermal expansion was measured using a thermo-mechanical analyzer. Preliminary observations seem to be in favourable agreement with the theoretical calculations.

ACKNOWLEDGEMENTS

The authors would like to thank Dr. A. Crasto of the University of Dayton Research Institute for his help in sample fabrication and Marlin Cook of Metcut Research for measurements of the coefficient of thermal expansion.

REFERENCES

(1) Hashin, Z., 1972, " Theory of Fiber Reinforced Materials ", NASA CR - 1974.
(2) Rosen, B. W., and Hashin, Z., 1970, " Effective Thermal Expansion Coefficients and Specific Heats of Composite Materials ", *International Journal of Engineering Science*, Vol. 8, pp. 157-173.
(3) Schapery, R. A., 1968, " Thermal Expansion Coefficients of Composite Materials Based on Energy Principles ", *Journal of Composite Materials*, Vol. 2, No. 3, pp. 380-404.
(4) Levin, V. M., 1967, " On the Coefficients of Thermal Expansion of Heterogeneous Materials ", *Mechanics of Solids*, Vol. 2, No. 1, pp. 58-61.
(5) Rosen, B. W., 1970, " Thermomechanical Properties of Fibrous Composites ", *Proceedings of the Royal Society*, London, Series A, Vol. 319, pp. 79-94.
(6) Hashin, Z., and Rosen, B. W., 1964, " The Elastic Moduli of Fiber Reinforced Materials ", *ASME Journal of Applied Mechanics*, Vol. 31, pp. 223-232.
(7) Chamis, C., C., and Sendeckyj, G. P., 1968, " Critique on Theories Predicting Thermoelastic Properties of Fibrous Composites ", *Journal of Composite Materials*, Vol. 2, No. 3, pp. 332-358.
(8) Hale, D. K., 1976, " Review : The Physical Properties of Composite Materials ", *Journal of Materials Science*, Vol. 11, pp. 2105-2141.
(9) Ishikawa, T., Koyama, K., and Kobayashi, S., 1978, " Thermal Expansion Coefficients of Unidirectional Composites ", *Journal of Composite Materials*, Vol. 12, pp. 153-168.
(10) Nomura, S., and Chou, T. W., 1981," Effective Thermoelastic Constants of Short-Fiber Composites ", *International Journal of Engineering Science*, Vol. 19, pp. 1-9.
(11) Takayo, Y., and Taya, M.,1985, "Thermal Expansion Coefficients and Thermal Stresses in an Aigned Short Fiber Composite With Application to a Short Carbon Fiber / Aluminum ", *ASME Journal of Applied Mechanics* ", Vol. 52, pp. 806-810.
(12) Takayo, Y., and Taya, M., 1987, " The Effect of Variable Fiber Aspect Ratio on the Stiffness and Thermal Expansion Coefficients of a Short Fiber Composite ", *Journal of Composite Materials*, Vol. 21, pp. 140-156.
(13) Schneider, W., 1971, *Kunstoffe*, Vol. 61, pp. 273.

(14) Yates, B., Overy, M. J., Sargent, J. P., McCalla, B. A., Kingston-Lee, D. M., Philips, L. N., and Rogers, K. F., 1978, " The Thermal Expansion of Carbon Fibre-Reinforced Plastics ",*Journal of Materials Science*, Vol. 13, pp. 433-440.
(15) Pagano, N. J., and Tandon, G. P., 1988, " Elastic Response of Multi - directional Coated-fiber Composites ", *Composite Science and Technology*, Vol. 31, pp. 273-293.

Table 1 : Selected thermo-mechanical properties of the fiber and matrix

Property Material	E (GPa)	ν	α $(10^{-6} / {}^{0}C)$
SCS -6 (Fiber)	413.7	0.24	2.4
H 3501-6 (Matrix)	4.27	0.34	51.15

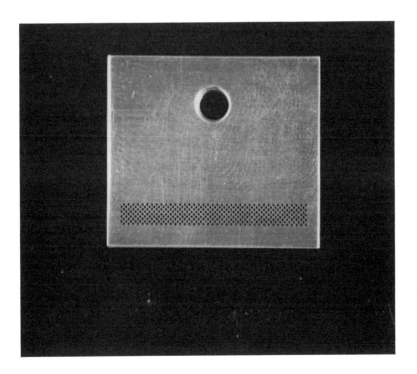

Fig 1   The aluminum jig with 360 holes drilled with a diameter of .375 mm and .625 mm apart

Fig 2   A macrophotograph showing the fibers in the jig before matrix infitration

Fig 3   A composite specimen with perfectly aligned fibers after matrix infiltration

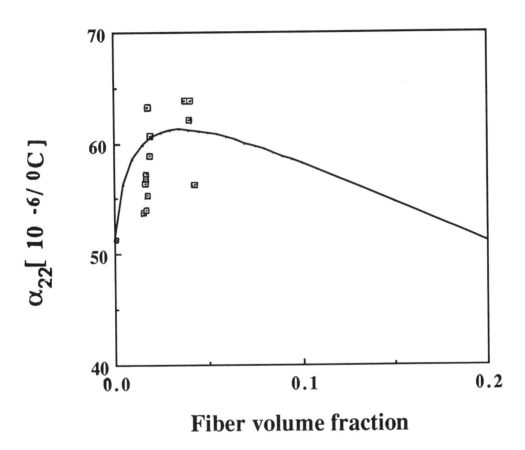

Fig 4 Transverse coefficient of thermal expansion of unidirectional SCS-6/H3501-6
composite system at various fiber volume fractions
(——— theoretical curve, ▣ experimental data points)

Thermal Expansion of Chemically Vapor Deposited Silicon Carbide Fibers

NEIL J. HILLMER

Abstract

Axial thermal expansion of CVD SiC fibers was measured from room temperature to 1500°C. The composition of these SiC fibers consists of a SiC sheath with a carbon-rich outer coating surrounding an unreacted graphite-coated carbon core. After a relatively small initial expansion from room temperature to 450°C, expansion was found to increase linearly with temperature from 450 to 1500°C. Above 1300°C a hysteresis effect was observed involving a temporary 50% reduction in expansion. Possible explanations for this hysteresis effect were considered and different theories are presented. Volume percent of carbon core was varied and found to have negligible effect on expansion. The conclusion was reached that expansion is controlled by the SiC sheath.

Introduction

The silicon carbide fibers examined in this study were produced by chemical vapor deposition (CVD) and are used for the reinforcement of metal matrix and ceramic composites. Specifically, these fibers are used in titanium and aluminum metal matrix composites because of their high-modulus, high-strength, thermal stability, and compatability with matrix materials. Ceramic matrix composites utilizing these CVD silicon carbide fibers exhibit increased fracture toughness.

N.J. Hillmer, Materials Engineer, Textron Specialty Materials 2 Industrial Ave, Lowell, MA 01851.

The objectives of this study were to confirm previous thermal expansion measurements made on these fibers and to generate data over a larger temperature range than obtained in past experiments [1,2]. The approach used was similar to that used by J. DiCarlo at NASA/Lewis and consisted of a constant axial load applied to a resistively heated fiber in an inert gas environment.

Accurate fiber expansion data is critical to the development of both metal and ceramic composites. In the case of metal matrix composites structural design modeling requires constituent fiber expansion data, while in the design of ceramic composites the thermal expansion characteristics of fiber and matrix should be matched to minimize stresses between the two.

Test specimens

The fibers examined in this study were produced by a chemical vapor deposition (CVD) process. The substrate for the process consists of a carbon monofilament. The carbon monofilament is resistively heated and coated with pyrolytic graphite prior to deposition of silicon carbide. The silicon carbide is deposited from the chemical decomposition of an organo-silane-hydrogen gas mixture [3].

The specific fibers examined here are the SCS-2, SCS-6 and SCS-9 silicon carbide fibers manufactured by Textron Specialty Materials, Lowell, MA. These fibers are differentiated by surface coatings and diameter. Figure 1 is a schematic representation of Auger plots indicating the ratio of Si/C within the surface coatings for these fibers. These fibers in themselves can be considered ceramic composites comprised of a carbon monofilament core and SiC sheath with a carbon-rich outer coating. The diameter of both the SCS-2 and SCS-6 fibers is nominally 142 microns, while the SCS-9 fiber has an outer diameter of only 76 microns. The diameter of the carbon core is ~37 microns for all three fibers.

It is important to note the crystal structure and grain growth pattern of the CVD SiC fibers. Warner et.al. [4,5] have shown that at the typical deposition temperature of 1350° C Beta-SiC with columnar grains oriented perpendicular to the substrate are produced. The preferred orientation of these cubic phase grains is <111>. There is also evidence of a hexagonal phase present with <001> preferred orientation. The columnar grains of the Beta-SiC are characterized by an extremely high density of (111) stacking faults and microtwins [5,6].

Test Procedure

The general methodology adopted for this study involved
suspending a weighted fiber with constant axial load applied
to it in an argon purged pyrex reactor, resistively heating
the fiber, monitoring temperature from 450° C --> 1500° C
with a series of optical pyrometers, and measuring fiber
deflection with a cathetometer. System qualification was
performed using well-characterized Molybdenum wire. Figure 2
illustrates the test set-up.

A Spellman power supply unit (Model No. RHP-3RW) was used to
resistively heat the weighted fiber held statically in a
pyrex tube 164 cm. in length. The temperature of the fiber
was adjusted by controlling the amperage output to the
fiber. To minimize oxidation effects, a high-purity Argon
purge was maintained throughout testing.

The temperature of the fiber was measured over the
temperature range 450-1500° C using three Williamson
dual-wavelength pyrometers, Model Nos. 8200, 8100, and 9100.
Each pyrometer covered a distinct range:

        No. 8200   450° C -- 850° C
        No. 8100   800° C -- 1100° C
        No. 9100   1000° C -- 1500° C

Pyrometer calibration was performed using a color-match
pyrometer at temperatures above 750° C.

Excellent agreement between color and dual-wavelength
pyrometers was found. Calibration was also confirmed by
accurate temperature overlap between the three pyrometers.

Resistance heating provided two important advantages over
other heating methods attempted: 1) a well-defined hot zone.
2) a relatively uniform temperature profile. Temperature
profile measurements were made along the length of the
fiber. A cold spot was observed at the gas inlet port.
Preheating the argon to >100° C reduced the cold spot to
less than 1/4" in length. Small temperature variations
(<30° at 1500° C) were observed elsewhere along the fiber
due to fiber diameter variations. During actual testing,
temperature measurements were made 79 cm. from the gas inlet
port.

The other critical variable of this study, deflection, was
measured using a Gaertner Scientific cathetometer. The
deflection of an 11.6 g. weight attached to the fiber was
measured as a function of temperature to a precision of
$\pm$ 0.05 mm.

System calibration was accomplished on a 36 micron diameter
Molybdenum wire. Measured expansion values were within $\pm 7\%$

of literature values at temperatures of 1000°C and 1250°C
[7]. The extremely small wire diameter precluded temperature
measurements below 1000°C and calibration at higher
temperatures was prohibited by creep effects.

Results

Figures 3-5 depict thermal expansion behavior of SCS-2,
SCS-6, and SCS-9, respectively. As can be seen from figures
3-5 there is a relatively small initial expansion from room
temperature to 450°C. Further studies are underway to
investigate the deviation from linearity over this
temperature range. Above 450°C, the fibers expand linearly
until 1300 - 1350°C (depending on the fiber).

At this temperature, a dramatic reduction in expansion is
observed with a gradual return to linear expansion at
~1470°C. Upon cooling the fiber, linear contraction
occurs until 1300-1350°C (again depending on the specific
SCS fiber) at which time the fiber does not change length
while cooling. Below 1200°C the fiber contracts linearly.
This hysteresis was observed on all three SCS fibers and was
also observed in previous studies [1,2] performed on SCS
fibers. Kern et. al. observed a similar hysteresis effect on
bulk CVD SiC. They attributed this phenomenon to the
contraction of free silicon when going from the solid to the
liquid phase at 1412°C [8]. Further discussion concerning
the existence of free silicon in these fibers will be
presented in the following section.

Discussion

The hysteresis effect observed for all three grades of SCS
fibers presents difficulties when incorporating the fibers in
a brittle matrix (i.e. ceramic composite). In almost all
ceramic matrices compatible with SCS fiber no hysteresis is
observed. Therefore, this effect can create thermal
expansion mismatch between matrix and fiber and thus produce
residual stress in the composites.

The exact cause of this hysteresis is still unclear. The
theory presented by Kern et. al. [8] attributing the
hysteresis to melting of free silicon could account for the
dramatic reduction in expansion with increasing temperature
at ~1300°C observed in this study, based on the fact that
silicon contracts in going from the solid to liquid phase.
The presence of free silicon in SCS fibers, however, has not
been proven definitively.

Evidence in support of the existence of free silicon has been
previously reported by DiCarlo in creep measurements of SCS
fibers. He attributed the observed creep in these fibers to
be due to grain boundary sliding (GBS) of Beta-SiC grains
[10]. It is generally accepted that the activation energy

for GBS should agree with either the lattice self-diffusion energy or the grain-boundary self-diffusion energy of the particular material, or with the lattice self-diffusion energy of the impurity phases in grain boundaries [9]. The creep activation energy measured by DiCarlo agreed favorably with the lattice self-diffusion energy of free silicon [10].

While DiCarlo's work suggests the presence of free silicon, previous auger and TEM examinations have not shown any unreacted silicon [6]. In addition, the presence of free silicon seems unlikely since an excess of carbon is maintained during manufacture of the fiber through the addition of excess amounts of propane gas.

In consideration of the absence of direct proof of free silicon in the fibers, other mechanisms might be causing the hysteresis. Analyzing the microstructure of the fiber one finds an extremely high density of stacking faults and microtwins [6]. It is possible that a temperature controlled morphological change occurs near 1300° C which could result in a reconfiguration of these stacking faults. Further investigation is necessary to substantiate this theory and is beyond the scope of this report.

One final point proven in this study is that the SiC sheath controls the expansion of the fiber. This fact is supported by good agreement between expansion values obtained for SCS fibers and those obtained for bulk Beta-SiC (Fig 3) [11]. This observation is anticipated due to the much greater elastic modulus of SiC (414 GPa) compared to that of the carbon core (34 GPa) [12,13]. In addition, the expansion behavior of SCS-6 and SCS-9, as can be seen in figures 4 and 5, is almost identical, even though the volume fractions of SiC and C are quite different. Thus indicating, the thermal expansion of SCS fibers is controlled by the SiC sheath and does not depend on the carbon-rich outer coating or carbon core.

Conclusion

The resistance heating technique developed in this study proved to be a viable test method for measuring thermal expansion of conductive fibers. Also proven in this study was that thermal expansion of SCS fibers is controlled by the SiC sheath and does not depend on the fiber coatings or the carbon core.

It is apparent from figures 3-5 that the thermal expansion behavior of SCS fibers is linear from 450 to 1300° C. Above this temperature all three fibers exhibited a hysteresis effect. Different theories explaining this phenomenon exist, such as the free silicon theory discussed. The absence of collaborating evidence of free silicon by powerful microstructural techniques like Auger analysis and TEM, allows for further speculation of possible causes of the hysteresis in SCS fibers.

References

[1] DiCarlo, J.A., "High Temperature Properties of CVD Silicon Carbide Fibers," Proceedings of an International Conference on Whisker and Fiber Toughened Ceramics, Oak Ridge National Laboratory, Oak Ridge, TN, 1988, pp 1-8.

[2] Brun, M., Communications of the American Ceramic Society in press.

[3] Debolt, H. and Krukonis, V, "Improvement of Manufacturing Methods for the Production of Low Cost Silicon Carbide Filament," Air Force Contract F33615-72-C-1177, Report No. AFML-TR-73-140, 1973.

[4] Wawner, F.E., Teng, A., and Nutt, S., "Microstructural Characterization of SiC (SCS) Filaments," Sampe Quarterly, 14 (3), Apr. 1983, pp 39-45.

[5] Nutt, S. and Wawner, F.E., "Silicon Carbide Filaments: Microstructure," J. Mat. Sci., Vol. 20, 1985.

[6] Wawner, F.E., "Boron and Silicon Carbide/Carbon Fibers", Fibre Reinforcements for Composite Materials, edited by A. R. Bunsell, Elseveir Science Publishers B.V., Amsterdam, 1988, pp 405-409.

[7] Metals Handbook, 9th edition, Vol. 6, ASM International, 1981, pg 772.

[8] Kern, E.L., Hamill, D.W., Diem, H.W., and Sheets, H.D., "Thermal Properties of Beta-Silicon Carbide from 20 to 2000°C," Mat. Res. Bull., Vol. 4, S25-S32, 1969.

[9] Nowick, A.S. and Berry, B.S., Anelastic Relaxation in Crystalline Solids, Academic Press, New York, 1972.

[10] DiCarlo, J.A., "Creep of CVD Silicon Carbide Fibers", J. Mater. Sci., Vol. 21, 1986, pp 217-224.

[11] Taylor, A. and Jones, R.M., "The Crystal Structure and Thermal Expansion of Cubic and Hexagonal Silicon Carbide," Proceedings of the Conference on Silicon Carbide, Boston, MA, 1959, pp 147-154.

[12] Giannetti, W.B., "Development of Elevated Temperature Test Techniques for Composites," Textron IRAD Report, Project #87009492, 1987, p. 1-59.

[13] Giannetti, W.B., "Sonic and Tensile Evaluation of Pitch Based Fibers," Textron Internal Report, July, 1985.

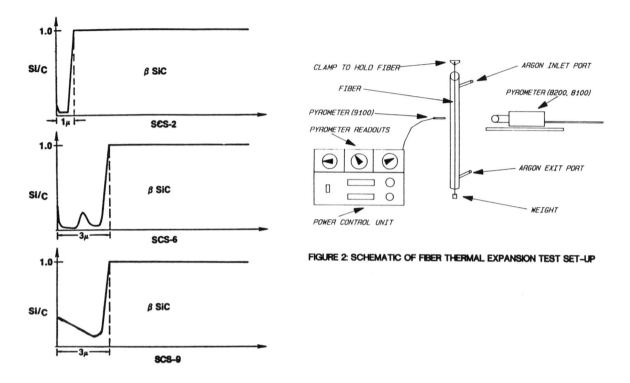

FIGURE 2: SCHEMATIC OF FIBER THERMAL EXPANSION TEST SET-UP

Figure 1: Schematic representation of Auger plots for SCS fibers

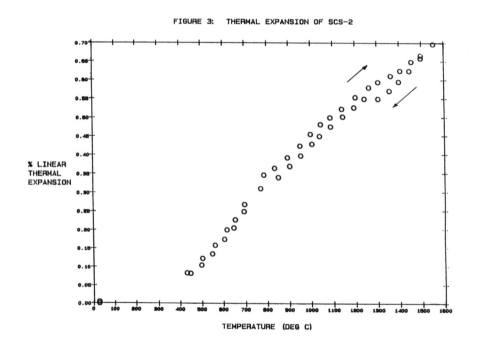

FIGURE 3: THERMAL EXPANSION OF SCS-2

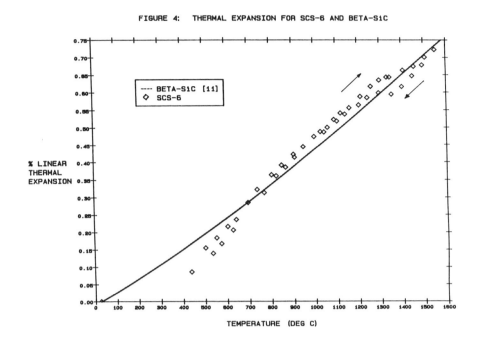

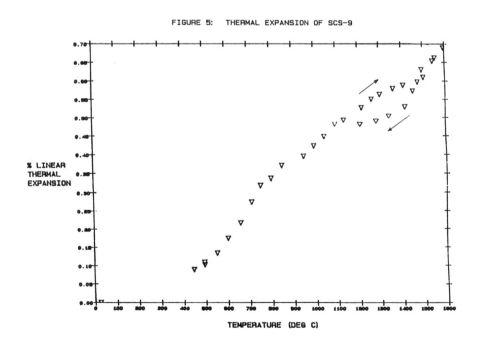

Size Effects in Carbon-Carbon Composites

R. A. HELLER, S. THANGJITHAM AND I. C. YEO

Paper withdrawn

Correlation of NDE Results on Metal Matrix and 2-D Carbon Carbon Composites

ROBERT W. PEPPER AND **THOMAS W. SHAHOOD, JR.**

ABSTRACT

The insurgence of high performance composites in the aerospace industry has prompted an accelerated effort to develop NDE techniques to characterize these materials. The task of the NDE engineer is not only to detect and characterize defects in these materials but also to make an attempt at correlating NDE data with properties.

This paper describes how NDE methods were used to characterize Metal Matrix and 2DCC materials to : identify the conditions detected by the various NDE methods; determine any overlap of sensitivity and capability of the various NDE methods; and describe the efforts in correlating NDE data with mechanical properties.

A comparison of NDE methods is discussed. NDE methods utilized in this study were Ultrasonics (C scans and velocity) and Radiography (High resolution film with opaque dye enhancement).

INTRODUCTION

The interest in fiber reinforced composite materials has been increasing with the advent of hypersonic aircraft concepts. Currently these composites are being evaluated using various nondestructive testing techniques. During the processing of these composites, many types of material variations and discrete discontinuities can be encountered which often can be traced directly to previous processing steps.

Robert Pepper, NDE Engineer
Thomas Shahood Jr., NDE Manager
Textron Specialty Materials
2 Industrial Avenue, Lowell, MA 01851

The NDE Engineer's conventional approach is to develop NDE techniques that will detect anomalies which may affect the intended performance of the composite parts. However, since little is known about the effects of defects on end use performance of these composite materials, the NDE Engineer must attempt to fully characterize the materials to provide design engineering with as much information as possible on material condition.

The effort described within was to explore x-ray and ultrasonic methodology for defect characterization in various composite materials. The task was then to correlate data generated by these methods and to verify defect size and type through destructive studies.

APPROACH

The approach was to evaluate samples of composite materials fabricated at Textron Specialty Materials (TSM). The configuration of the samples and the testing performed are as follows :

Seeded Defect Panel #1:
In this series the composite material consisted of titanium 15-3-3 matrix reinforced with TSM's continuous silicon carbide fiber. (SCS-6)

The concept of a "seeded defect" panel is well known in the composite NDE community. The approach is to implant foreign materials or process anomalies into the composite in a controlled manner to simulate defects that may be encountered in the actual material.

TSM fabricated a metal matrix composite panel with known defects implanted in the center zero degree ply. The implanted defects included fiber distortions, fiber breaks and boron nitride seeded areas (to simulate poor consolidation). In the seeded defect panel used for this study the fiber spacing was 126 fibers per inch with additional sections containing 124,135 and 145 fibers per inch. (See Fig. 1).

Conventional Ultrasonic C-Scan inspection was performed on the panel. A through transmission technique was utilized at 10 Mhz and a continuous grey scale recording was produced. The ultrasonic instrument utilized was an Automation Industries model M91-C. The reference plate used was a piece of titanium 15-3 sheet stock of a thickness equivalent to the plate with .005" thick lead tape disks 1/8" and 1/4" in diameter attached to the plate. The purpose of the reference plate is to establish a repeatable base line for instrument gain settings and to verify alignment of the ultrasonic beam. All subsequent scans are described, in terms of gain setting, as the amount of instrument gain (in Db) above or below the setting required to obtain 100% screen height on the reference plate.

The panel was then radiographed using high resolution film radiography. This technique utilizes extra fine grain film (Dupont NDT-30), low kilovoltage potentials and relatively high Miliamperages. The resultant radiographic images are then viewed utilizing a high intensity film viewer coupled to a stereo zoom microscope. This technique results in magnified, high resolution images of the composite. Polaroid or 35 mm photos can then be generated through the microscope to create hard copies of the magnified radiographic images.

The panel was then sectioned for correlation of NDT results. Macro and microphotography were utilized to evaluate the edges of the specimens. The specimens were then subjected to a 10 to 30 minute soak in the x-ray opaque dye Tetrabromoethane (TBE) and subsequently radiographed to trace any laminar flaws or interconnected porosity sites that may have been exposed after machining.

Titanium Metal Matrix Composite Panels:

A panel from TSM's on going R+D programs in Titanium metal matrix composite programs was used for this study. This panel (TSM 86H037) was fabricated as part of a study for evaluating tensile properties at elevated temperatures. The panel was 6 ply unidirectional silicon carbide filament (TSM SCS-6) in Titanium 15-3-3 matrix.

NDE consisted of Ultrasonic c-scan inspection performed as described previously; ultrasonic velocity measurements and by High resolution radiography utilizing methods described previously.

The panel was further processed into tensile specimens for their intended use. The machined specimens were subjected to Macro and microphotography as well as opaque dye enhanced radiography to detect any exposed defects.

Miscellaneous Titanium Metal Matrix Composite Parts

Various Titanium metal matrix components were evaluated in process. These parts were subjected to ultrasonic and radiographic testing as well as some destructive correlation. Since these parts were part of the normal production flow, the thoroughness of analysis performed was subject to internal priorities and delivery schedules.

2-D Thin Wall Carbon / Carbon Composite Development

The material selected for this study was 2DCC eight harness satin weave. The material was processed from carbon 641 /T300 HT prepreg yarn. 15 plies were molded in a 0-90 degree orientation with a 50 % fiber volume and a SC-1008 phenolic resin. The panel dimensions, weight, volume, fiber volume and density were recorded throughout the four cycles of carbonization, graphitization and impregnation.

Satin weave fabrics are very pliable and conform readily to complex shapes which can be woven into high densities offering offering high strength in both directions. Because of these features, thin walled carbon/carbon materials have potential application for skin and structural components on Hypersonic Aerospace vehicles. Several panels were processed in an effort to develop NDE methods which would aid in defining integrity while providing insight on performance.
NDE techniques used were similar to those utilized to evaluate the metal matrix composites with the addition of computer aided ultrasonics. The NDE data was used to lay out specimens for tensile and short beam shear tests. The specimens were evaluated using conventional microscopy, scanning electron microscopy and metallographic methods.

TEST RESULTS (Reference Figures 1 through 20)

Titanium Metal Matrix Seeded Defect Panel

Figure 1B shows the ultrasonic c-scan results on the seeded defect panel. The scan represents a gain level of 6 Db above 100% on the reference plate. The implanted Boron Nitride inclusions appear to have migrated toward the central portion of the panel contributing additional poor consolidation indications. The three oval shaped areas (cut patches) are visible on the scan, where fiber breaks occurred. The cut fibers are not visible nor are the varied fiber spacing sites. Four samples (1 through 4) were removed from the panel as shown on the scan.
Figure 2A shows a radiographic image of the 1/2" cut oval patch in the fiber. Note the clearly distinguishable fiber terminations.
Figure 2B shows a opaque dye enhanced radiographic image of the same area. The close correlation with c-scan

data demonstrates the ability of dye enhanced
radiography to detect laminar flaws.
Figure 3A+B show photomacrographs of the 1/2" cut patch
defect. Note the inadequate metal flow on the top foil
between parallel fibers and the disrupted 2nd foil
interface with the middle fiber ply. This defect was
not intentional. It does explain why the c-scan data
and the dye enhanced radiographic data indicated laminar
flaws.
Figure 4A+B shows a comparison of conventional and
opaque dye enhanced radiographs of the zone of major
ultrasonic attenuation from c-scan data (specimen 4).
The dye enhanced image shows almost complete dye
migration in the area of poor sound transmission. The
top left corner of the specimen shows no dye migration
which correlates with c-scan data which indicates good
sound transmission.

Titanium Metal Matrix R+D Panels

Figure 5A+B shows c-scan and velocity data from panel
86H037. The c-scan shows poor sound transmission in zone
A and good sound transmission in zone B. No significant
variation was seen in velocity or conventional
radiographic evaluation. The panel was sectioned
parallel to fiber lay up and specimens were removed from
areas of good and poor sound transmission.
Figures 6A+B are photomicrographs taken from the zone A
region of poor ultrasonic transmission. Inadequate
consolidation, porosity sites, bunched fibers,and large
grain sizes are seen.
Figure 7A+B are photomicrographs taken from the zone B
region of good ultrasonic transmission. Note the more
uniform fiber spacing and the absence of porosity.

Miscellaneous Metal Matrix Parts

Figure 8A-D compares the inspection results of an eight
ply titanium metal matrix panel with SCS-6 fiber and
titanium 6Al/4V matrix. A processing problem resulted
in surface blistering with associated surface cracks.

Figure 8A : Visual inspection of the panel at 10X
magnification
Figure 8B : Ultrasonic c-scan of same defect at 50
Mhz using a pulsed echo technique. This scan revealed
surface cracks as well as poor consolidation.
Penetration of the sound beam was minimal due to the
high frequency rendering the technique applicable only
to surface flaws.
Figure 8C : Macroradiograph of the defect after a 10
minute soak in radiopaque dye. All open to the surface
cracks allow dye migration between ply layers.

<u>Figure 8D</u> : C-scan image using a 10 Mhz focused transducer indicating gross delamination. These conventional c-scan techniques are not sensitive to fiber breaks or surface cracking.

<u>Figure 9</u> compares a high resolution radiograph with a computerized ultrasonic c-scan on a 45 degree implanted SCS-6 fiber patch in a 0-90 degree lay up. The percent of sound loss was approximately 20 Db at 10 Mhz indicating poor consolidation.

<u>Figure 10</u> is a comparison of high resolution radiography and computerized ultrasonic c-scan on a fiber distortion commonly identified as a "fish eye". The ultrasonic signal loss from this condition varied from 7 to 15 Db at 25 Mhz.

<u>Figure 11</u> is a comparison of high resolution radiography and computerized ultrasonic c-scan of a single ply cut in an 8 ply panel. Note the small gap where the ply cut occurred. The ultrasonic signal loss on this defect was -4Db at 25 Mhz.

<u>Figure 12</u> is a comparison of high resolution radiography and computerized ultrasonic c-scan of a fabric overlap. This condition is very difficult to resolve radiographically due to the directionality of the fibers. Ultrasonic c-scan inspection clearly detected this condition as a 40 Db signal loss at 25 Mhz.

<u>Figure 13</u> is a comparison of high resolution radiography and computerized ultrasonic c-scan of a implanted Yiitrium Oxide spot centered in an eight ply panel. Note the high x-ray absorption of the spot on the radiograph. The ultrasonic signal loss from this indication was approximately 40 Db at 25 Mhz.

<u>Figure 14</u> shows high resolution radiographs of implanted broken fiber defects. This is a dramatic example of the ability of high resolution radiographs to distinguish fiber defects.

2D Thin Wall Carbon / Carbon Composite

<u>Figure 15</u> shows an attenuation c-scan using a reflector plate pulse echo technique and a 5 Mhz focused transducer. Approximately 60% of the panel exhibited a 40 Db signal loss from the remainder of the panel.

<u>Figure 16</u> is a c-scan image from two tensile specimens from panel 111 using the same set up as in Fig. 15. One specimen was taken from the area of poor ultrasonic transmission and the other was taken from an area of good ultrasonic transmission. The specimens showed no sign of degradation due to machining and the c-scan results correlated closely with the original panel c-scan results.

<u>Figure 17A+B</u> are photomacro and micrographs of tensile specimen B which revealed poor ultrasonic transmission.

A striation is visible along the length of the overlapping fibers.

Figure 18 shows macroradiographic images of specimen B using x-ray opaque dye. These images offer better information on the on the nature of planer defects as compared to the other NDE methods utilized.

Figure 19A+B are macroradiographs of tensile specimen B before and after treatment with the opaque dye. Interlaminar discontinuities such as voids and delaminations are not readily discernible by radiography unless said defects are enhanced by opaque dye chemistries.

Figure 20 is a photomicrograph of a resin rich agglomeration taken of the machined edge of specimen A which produced a high ultrasonic signal response. The scanning electron microscope clearly indicated the accumulation of phenolic resin in the outer ply layer.

Figure 21A+B,C and D shows the processing and mechanical test data for the 2DCC material. No correlation between the individual mechanical test data and the NDE data found by this study.

Conclusions

Magnified High resolution film radiography combined with ultrasonic inspection techniques have shown to be successful in detecting and documenting various anomalies in these composite materials. The role of both NDE methods have shown to be complementary to each other in the full characterization of the materials. Although radiography is successful in determining fiber integrity and lay up anomalies, it falls short of detecting defects related to processing such as porosity, lack of consolidation, lack of cure and delamination. Ultrasonics, on the other hand has proven to be successful in detecting processing defects such as porosity, delamination and lack of consolidation or cure but we have shown that the method falls short of detecting fiber related defects such as breakage and distortion.

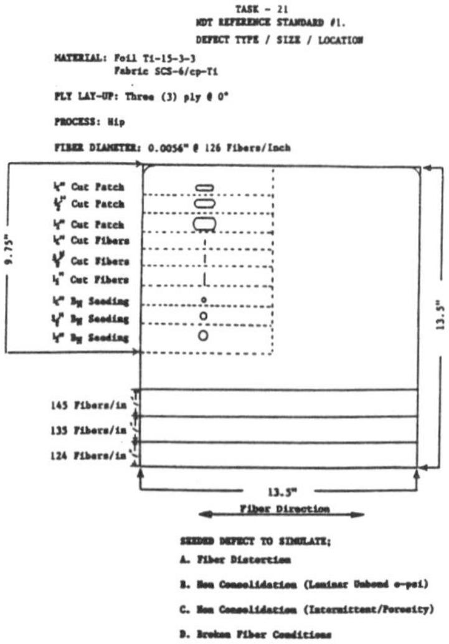

TASK - 21
NDT REFERENCE STANDARD #1.
DEFECT TYPE / SIZE / LOCATION

MATERIAL: Foil Ti-15-3-3
Fabric SCS-6/cp-Ti

PLY LAY-UP: Three (3) ply @ 0°

PROCESS: Hip

FIBER DIAMETER: 0.0056" @ 126 Fibers/Inch

⅛" Cut Patch
¼" Cut Patch
½" Cut Patch
⅛" Cut Fibers
¼" Cut Fibers
½" Cut Fibers
⅛" Dry Seeding
¼" Dry Seeding
½" Dry Seeding

9.75"
13.5"

145 Fibers/in
135 Fibers/in
124 Fibers/in

13.5"

Fiber Direction

SEEDED DEFECT TO SIMULATE:

A. Fiber Distortion

B. Non Consolidation (Laminar Unbond ε-psi)

C. Non Consolidation (Intermittent/Porosity)

D. Broken Fiber Conditions

Fig. 1A

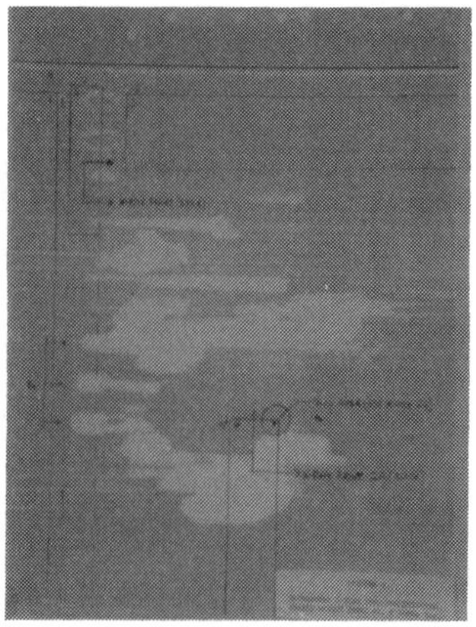

Fig 1B

Fig 2A

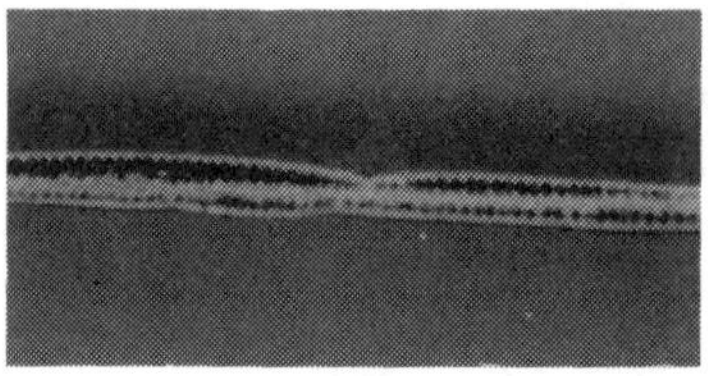

Fig 3A

Fig 4A

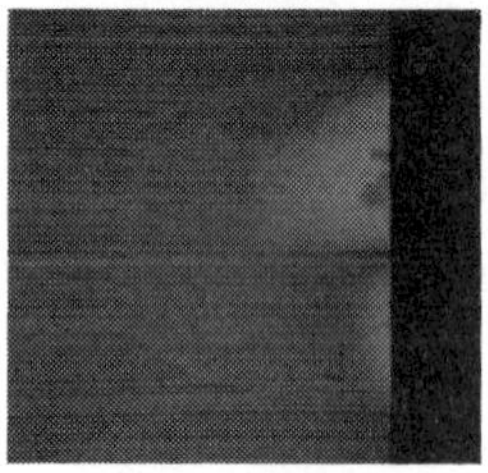

Fig 2B

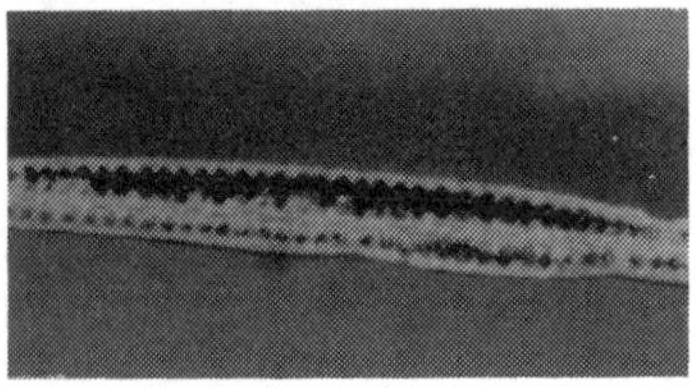

Fig 3B

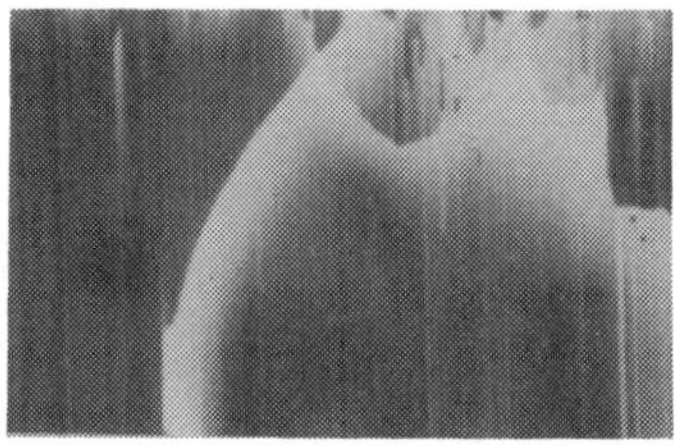

Fig 4B

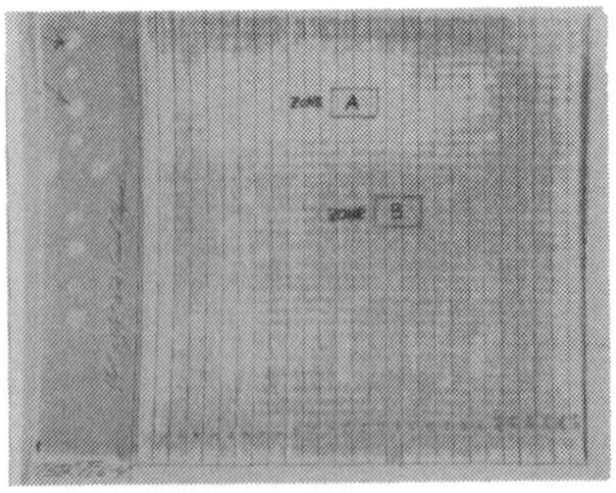

Fig 5A

	Long Time (usec)	Shear Time (usec)	Sample Thickness (in)	Sample Density (gm/cc)	Long Vel (in/sec)	Shear Vel (in/sec)	Poisson's Ratio	Young's Modulus (PSI 10^6)
1	0.251	0.477	0.0370	3.610	294821	155136	0.3085	21.267
2	0.253	0.485	0.0375	3.610	296443	154639	0.3131	21.204
3	0.259	0.480	0.0379	3.610	292664	157917	0.2946	21.802
4	0.254	0.476	0.0376	3.610	296063	157983	0.3010	21.927
5	0.247	0.476	0.0370	3.610	299595	155462	0.3158	21.474
6	0.258	0.482	0.0371	3.610	287597	153942	0.2992	20.792
7	0.250	0.484	0.0372	3.610	297600	153719	0.3181	21.032
8	0.255	0.499	0.0377	3.610	295686	151102	0.3233	20.403
9	0.265	0.499	0.0374	3.610	282264	149900	0.3036	19.781
AVG	0.255	0.484	0.0374	3.610	293637	154422	0.3086	21.076

Fig 5B

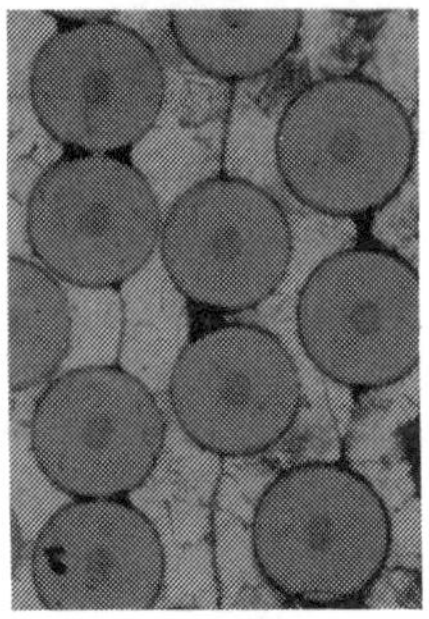

Fig 6A

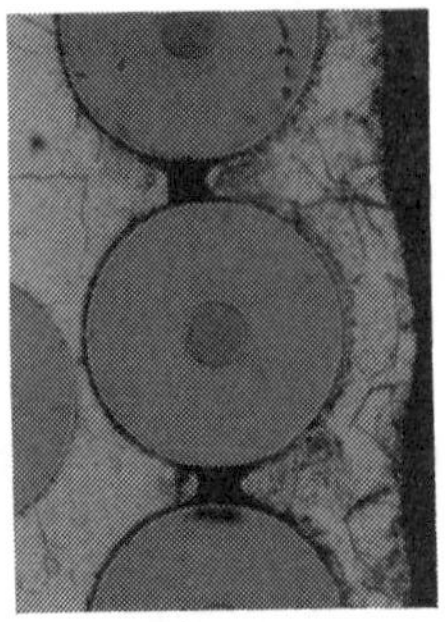

Fig 6B

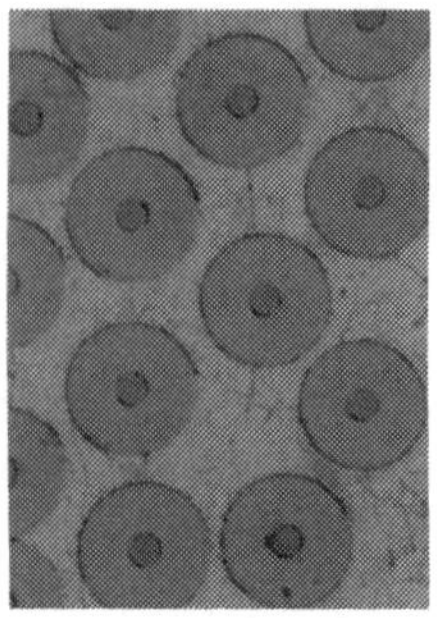

Fig 7A

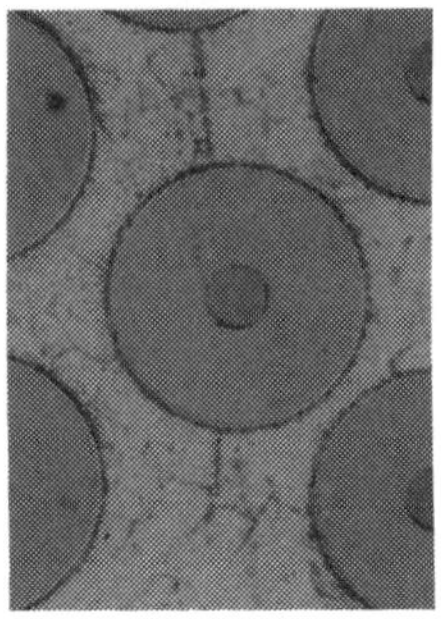

Fig 7B

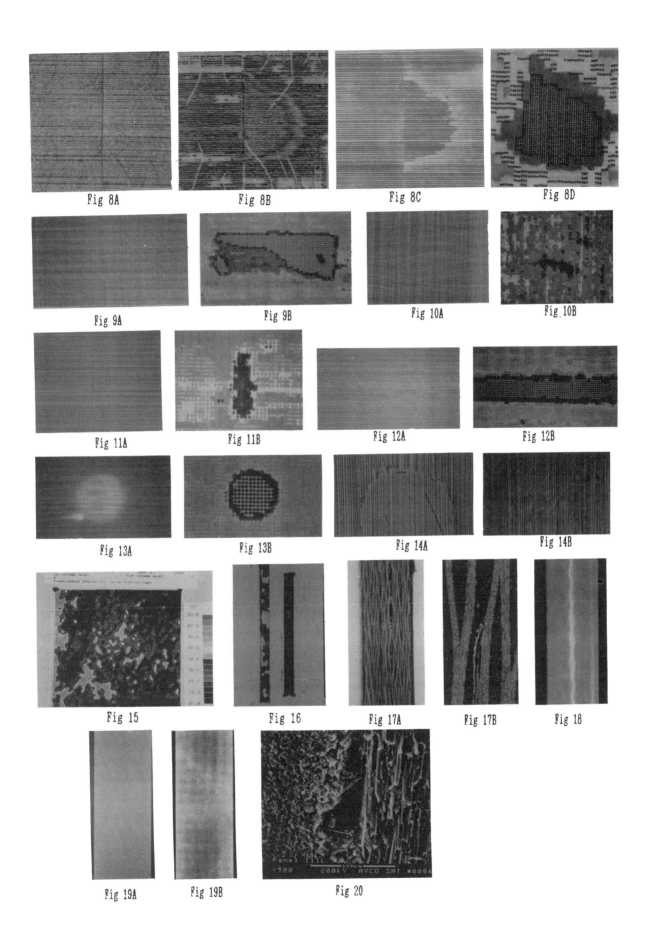

Fig 8A

Fig 8B

Fig 8C

Fig 8D

Fig 9A

Fig 9B

Fig 10A

Fig 10B

Fig 11A

Fig 11B

Fig 12A

Fig 12B

Fig 13A

Fig 13B

Fig 14A

Fig 14B

Fig 15

Fig 16

Fig 17A

Fig 17B

Fig 18

Fig 19A

Fig 19B

Fig 20

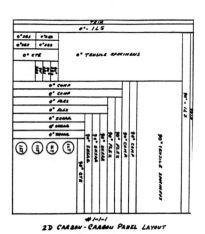

Fig 21A

2D CARBON-CARBON PANEL LAYOUT

PROGRAM						
2D - THIN WALL CARBON-CARBON DEVELOPMENT						
TYPE OF PREPREG/YARN				TYPE OF RESIN		
KARBON 641 / T300 HT				SC1008 PHENOLIC		
No. OF PLIES	ORIENTATION OF PLIES			PANEL No.		
15	0°, 90°			1-1-1		
MODE OF FABRICATION						
AUTOCLAVE						

PROCESS STEP	AVERAGE LENGTH	AVERAGE WIDTH	AVERAGE THICKNESS	WEIGHT	VOLUME	FIBER VOLUME	DENSITY ρ
MOLD	11.581"	11.773"	.200"	696.5g	446.89CM³	.514	1.56G/CM³
CARB #1	11.716"	11.585"	.186"	588.0g	413.70CM³	.553	1.421G/CM³
GRAPH #1	11.717"	11.582"	.185"	578.8g	411.41CM³	.556	1.407G/CM³
IMPREG #1 CARB #2	11.729"	11.600"	.184"	638.0g	410.24CM³	.559	1.555G/CM³
IMPREG/ CARB 3	11.735"	11.586"	.182"	678.0g	405.50CM³	.565	1.672G/CM³
GRAPH 2	11.719"	11.588"	.182"	670.5g	405.01CM³	.565	1.656G/CM³
IMPREG/ CARB 4	11.720"	11.581"	.182"	705.5g	404.81CM³	.565	1.743G/CM³

Fig 21B

T-300 CONTINUOUS YARN
8 HARNESS SATIN WEAVE - KARBON 641
PANEL # 1-1-1

PROCESS STEP	V/O FIBER	V/O RESIN CHAR	V/O VOID
MOLD	.514	.432	.054
CARB #1	.553	.181	.266
GRAPH #1	.556	.159	.285
IMPREG. CARB #2	.559	.251	.190
IMPREG. CARB #3	.565	.312	.123
GRAPH #2	.565	.286	.149
IMPREG. 3/CARB 4	.565	.353	.082

Fig 21C

2D C/C Panel III

	Table X Short Beam Shear		Table Y Tensile Properties		
	0°	90°	Strength (psi)	Modulus (x 10⁶ psi)	Total Strain (%)
x	1326	1322	43,750	16.3	.39
s	58	19	300	0.3	.03
cv	4%	1.4%	0.6%	1.8%	8%

The test data in Tables X & Y indicate a material that has uniform properties in each direction in spite of an ultrasonic mapping that indicated significant variations within the panel. Additional testing (edgewise compression) may be necessary to verify any effects of in plane anomolies.

Fig 21D

Investigation of Test Methods for Measuring Compressive Strength and Modulus of Two-Dimensional Carbon-Carbon Composites

CRAIG W. OHLHORST, JAMES WAYNE SAWYER AND Y. ROBERT YAMAKI

ABSTRACT

Carbon-carbon composite materials are being considered for use in high temperature structural applications where accurate mechanical property data are essential for design purposes. To date, no standard test methods have been adopted for measuring mechanical properties of carbon-carbon composites. An experimental program has been conducted to investigate two test techniques for measuring in-plane compressive failure strength and modulus values for both coated and uncoated ACC-4 type carbon-carbon material. Test techniques investigated consisted of testing specimens with potted ends and testing specimens in a newly designed clamping fixture. Test parameters investigated were specimen shape, length, gauge width, and thickness. Measured compressive failure strength and modulus values are presented for coated and uncoated 0/90° and ±45° laminates. Test results show that specimen shape does not have a significant effect on the measured compressive properties but specimen gauge widths below 0.50 inches should be avoided. The effect of specimen length still remains to be resolved. Potting the ends of the specimens results in slightly higher measured compressive strengths than obtained with the new clamping fixture but comparable modulus values are obtained using both techniques.

INTRODUCTION

Carbon-carbon composite materials have excellent weight-strength properties at very high temperatures and are being considered for use in various high temperature structural applications. Accurate mechanical strength and modulus values are essential for design purposes. To date, however, no standard test methods have been adopted for measuring mechanical properties of carbon-carbon composites. Test specimen configurations and test techniques for measuring the interlaminar tensile and shear strength of carbon-carbon materials were studied in the investigations reported in references [1] and [2], respectively. Both of these reports dealt with interlaminar properties which are weak in two-dimensional carbon-carbon composite materials and must be accounted for in designing structural components. Accurate in-plane mechanical strength and modulus values are also essential for design purposes as was shown in reference [3] where compression modulus was found to have a significant effect on the calculated buckling loads for blade stiffened panels. Thus, a test technique is needed to accurately measure the compressive strength and modulus of carbon-carbon materials.

ASTM D695-80 [4] is commonly used to measure compressive strength for composites even though it was developed for rigid plastics. One drawback of this technique is that compressive modulus can not be measured since there is no way to attach an extensometer.

Craig W. Ohlhorst and James Wayne Sawyer, NASA Langley Research Center, Hampton, Virginia, Y. Robert Yamaki, Planning Research Corporation, Hampton, Virginia

233

This paper describes a parametric investigation of test methodology for measuring compressive strength and modulus for both coated and uncoated ACC-4 type carbon-carbon material. Test techniques investigated consisted of testing some specimens with potted ends and others using a test fixture designed and fabricated at the NASA Langley Research Center. Test parameters investigated include specimen shape, length, and gage width. Compressive failure strength and modulus values will be presented and discussed for both coated and uncoated carbon-carbon material.

MATERIALS

The uncoated specimen material was ACC-4 type sheet material supplied by Carbon-Carbon Advanced Technologies, Inc.* and contained 9 plies of heat-treated T300, eight-harness satin weave fabric having 24 warp ends per inch by 23 fill ends per inch. The plies were laid up with alternate 0 and 90° orientations of the fabric warp direction. After the final pyrolysis step, the material was given an inert heat treatment to simulate a coating cycle.

The coated ACC-4 sheet material was supplied by Vought Corporation. It was fabricated using 6 and 12 plies of heat-treated T300, eight-harness satin weave fabric. It was laid up the same way as the uncoated material. The substrate was conversion coated to a depth of approximately 0.015 inches with silicon carbide followed by a tetraethyl orthosilicate high-temperature sealer and an additional low-temperature sealer. Further processing details for the material are given in reference [5].

TESTS

Compressive strength and modulus measurements were made for the carbon-carbon materials using two test techniques, two specimen shape configurations, and specimens with various length, gauge width, and thickness dimensions. Specimens were cut from the sheet material using water cooled, diamond impregnated cutting tools. Some specimens were cut with the length dimension parallel and perpendicular to the fiber direction (0/90° laminate) while others were cut with the length dimension ± 45° to the fiber directions (±45° laminate). A summary of the tests conducted are listed in Table I through IV where the test technique, specimen configuration, and specimen dimensions are given.

Tests were conducted to evaluate two techniques and to determine how sensitive the measured strength and modulus values are to the test technique used. One technique consisted of potting both ends of the specimen with epoxy potting compound as shown in figure 1. The ends of the specimens were ground flat and parallel and the platens of the test machine were adjusted to be parallel to insure uniform loading of the specimen. Tests were conducted by applying a load to the specimen by direct contact with the two platens of the test machine. The second technique investigated used a test fixture designed and fabricated at NASA Langley Research Center (fig. 2). The fixture consists primarily of two support plates between which the specimen is lightly clamped by means of screws. In addition, shallow steps on the internal faces of the plates are mated with the specimen edges by means of lateral adjustment screws. All screws are tightened finger tight to allow the supported and aligned specimen to slide easily within the fixture. A central opening is machined through both plates to allow attachment of an extensometer to the specimen, and to permit specimen failure without the constraining effects of the support plates. The specimens are cut nominally 0.05 inches longer than the test fixture so that the specimens can be loaded without compressing the fixture. Tests are conducted by mounting the specimen and fixture in the test frame between two self-leveling platens as shown in figure 2 to help insure alignment of the specimen with the load train.

The two specimen shape configurations tested are shown in figure 3. Configuration A is a dog bone shape with the base 0.25 inches wider than the gauge section and with a 0.75

---

* Identification of commercial products does not constitute endorsement, expressed or implied, of such products by the National Aeronautics and Space Administration.

inch shoulder radius as shown. The gauge length varied as a function of specimen length. Configuration B is a rectangular shape. The two shape configurations were investigated to determine if specimen shape has a significant effect on the measured strength and modulus values and to determine if the added time required to machine the dog bone shape is necessary.

Tests were conducted on specimens of various lengths and widths to determine the specimen dimensions required to obtain accurate measurements of compression strength and modulus. Tests were conducted for specimen lengths of 2.55 and 3.05 inches for the uncoated specimens and for 3.05 inches for the coated specimens. Specimen widths of 0.25, 0.50 and 0.75 inches were tested for both the uncoated and coated specimens. The uncoated 9-ply material had a nominal thickness of 0.105 inches. The coated 6-ply material had a nominal thickness of 0.074 inches while the coated 12-ply material had a nominal thickness of 0.138 inches. Two specimen thicknesses were chosen for the coated material tests because it was expected that the coating would have a larger effect on the test results for the thinner 6-ply material than for the thicker 12-ply material and thus could affect the specimen dimensions required to obtain accurate compression strength and modulus values.

A universal screw driven test machine was used to test the specimens. All tests were conducted at a constant cross-head displacement rate of 0.02 in/min. A clip gauge extensometer was attached to each of the specimens to measure the displacement over a 1.00 inch section of the specimen. The applied load and specimen displacement were recorded on an x-y plotter.

RESULTS AND DISCUSSION

Compressive strength and modulus values for all specimens tested are listed in tables I through IV. Values given for the coated material are based on total specimen thickness including coating. For the uncoated 0/90° laminate specimens the strength values ranged from 23.4 to 26.3 ksi and the modulus values ranged from 14.9 to 20.8 Msi. For the coated 0/90° laminate specimens, the strength and modulus values obtained for both the 6- and 12-ply material were almost identical to those obtained for the uncoated 0/90° laminate specimens. Slightly more scatter in the data was obtained for the coated material than for the uncoated material. However, in general, coating the 0/90° laminate specimens had little effect on either the compressive strength or modulus values.

For the uncoated ±45° laminate specimens, the strength values ranged from 7.5 to 8.3 ksi and the modulus values ranged from 2.9 to 3.4 Msi. For the coated ±45° laminate specimens, the strength values ranged from 14.0 to 16.3 ksi and the modulus values ranged from 4.9 to 6.2 Msi. Thus coating the ±45° laminate specimens almost doubles both the strength and modulus values. Similar strength and modulus values were obtained for both the 6- and 12-ply coated material. The large increase in strength and modulus values is probably due to the contribution from the high modulus (70 Msi) silicon carbide coating [6], which is considerably higher than that for the ±45° laminate substrate material. Although the modulus value for the silicon carbide coating is also higher than that for the 0/90° laminate substrate, the difference is not nearly as great as for the ±45° laminate.

The effect of specimen length was studied using the uncoated 0/90° laminate specimens and the potted end test technique. Compressive strength and modulus values are shown in figure 4 for the 2.55 and 3.05 inch long specimens. Maximum, minimum, and average values are shown by the three horizontal lines at the top of the bars and the numbers at the top of the bars indicate the number of replicate specimens tested for each configuration. Strength and modulus values were slightly lower for the 3.05 inch long specimens than for the 2.55 inch long specimens. The reason for this difference is not presently understood, and further work is needed to resolve this issue.

Strength and modulus values measured for the uncoated 0/90° laminate specimens using the two test techniques are shown in figure 5 for both the dog bone (config. A) and rectangular (config. B) specimens. The modulus values are almost identical using both test techniques and both test specimen configurations. Similarly, the difference in shapes (dog bone and rectangular), had little effect on the measured strength values. However, the strength values measured using the potted end technique are slightly higher than those obtained using the clamped end fixture. The difference is probably due to a more uniform introduction of the load into the carbon-carbon material with the potted ends than with the clamped end fixture.

The effect of specimen gauge width on the measured compressive strength and modulus values for uncoated material is shown in figures 6 and 7 for the 0/90° and ±45° laminate specimens, respectively. For the 0/90° laminate specimens (fig. 6) results are shown for both dog bone and rectangular cut specimens. Similar results were obtained using both shape configurations. For specimen gauge widths of 0.50 and 0.75 inches, the gauge width did not have a significant effect on the measured strength and modulus values. For the 0.25 inch gauge width specimens, the strength again was not affected but the modulus values are suspect. Two of the modulus values were close to 20 Msi. It is suspected that due to the narrow gauge width some extensometer slippage might have occurred leading to higher modulus values. For the ±45° laminate specimens (fig. 7), gauge width (0.25 - 0.75 inches) did not have any significant effect on either strength or modulus values.

Compressive strength and modulus values measured for the coated 0/90° laminate specimens are shown in figure 8 for both the 6- and 12-ply materials. Test results are shown for specimens gauge widths of 0.25, 0.50, and 0.75 inches. Comparison of the results for the different widths show that the gauge width has no significant effect on the measured strength and modulus values. The 12-ply material has higher strength and modulus values than does the 6-ply material. This increase in strength and modulus values for the thicker material is probably due to a slightly higher fiber percentage for the thicker material. The higher fiber percentage for the 12-ply material is indicated by the different thickness per ply for the 12-ply material (0.0115 inches calculated from table III) as compared with the 6-ply material (0.0123 inches).

Compressive strength and modulus values measured for the coated ±45° laminate specimens are shown in figure 9 for both the 6- and 12-ply materials and for gauge widths of 0.25, 0.50, and 0.75 inches. For both the 6- and 12-ply materials, the measured compressive strength and modulus values are slightly lower for the 0.25 inch gauge width specimens than for the wider specimens. Thus, it appears that for the ±45° laminate specimens, widths greater than 0.25 inches are required to give more reliable compressive strength and modulus values. The strength and modulus values obtained for the 0.50 and 0.75 inch gauge width specimens are approximately equal which indicates that a specimen width of 0.50 inches is adequate.

Comparison of results in figure 9 for the two material thicknesses show that the strength values are higher and the modulus values lower for the 6-ply material than for the 12-ply material. As was pointed out previously, the coating significantly increases the compressive strength and modulus values for both thicknesses of the ±45° laminate specimens. Since the coating thickness is approximately the same for both substrate thicknesses, the coating comprised a larger percentage of the material thickness for the 6-ply material than for the 12-ply material. Thus, the higher strength values for the 6-ply material than for the 12-ply material are understandable. However, the lower modulus values for the 6-ply than for the 12-ply material remain to be explained.

CONCLUDING REMARKS AND RECOMMENDATIONS

An experimental investigation has been conducted on two test techniques for measuring in-plane compressive strength and modulus values for both coated and uncoated ACC-4 type carbon-carbon material. Test techniques investigated consisted of specimens with potted ends and specimens clamped in a specially designed clamping fixture. Test parameters investigated were specimen shape, length, and gauge width. Tests were conducted on coated and uncoated specimens cut parallel and perpendicular (0/90° laminate) to the warp direction of the layup and on specimens cut ±45° to the warp direction of the layup (±45° laminates). Uncoated specimens were 9-plies thick and the coated specimens were 6- and 12-plies thick.

Test results for the uncoated material showed that a specimen length of 3.05 inches gave slightly lower strength and modulus values than obtained from specimens 2.55 inches long. The reason for this difference is not clear at this time. Pending resolution of this issue, an interim recommendation is to test at the 2.55 inch length. Compressive strength and modulus values obtained with dog bone shaped specimens were not significantly different from values obtained with rectangular shaped specimens. Tests on specimens with potted ends resulted in slightly higher measured strength values but approximately equal modulus values in comparison with specimens tested in the specially designed fixture.

Test results for both coated and uncoated materials showed that specimen widths of 0.50 inches or greater did not have a significant effect on the measured compressive strength and modulus values. For the 0.25 inch wide specimens, measurement difficulties were encountered for the uncoated 0/90° laminates. For the coated ±45° laminates, lower strength and modulus values were obtained for the 0.25 inch wide specimens than for wider specimens. Thus, specimens 0.50 inches or wider are recommended for measurement of in-plane compressive properties.

Comparison of the test results for the coated and uncoated material shows that applying a silicon carbide conversion coating to the carbon-carbon substrate does not have a significant effect on the compressive strength and modulus values of the 0/90° laminate material. However, for the ±45° laminate material, a conversion coating almost doubles the compressive strength and modulus values. Specimen thickness has a small effect on the compressive strength and modulus values.

REFERENCES

1.   Ransone, Philip O., Maahs, Howard G., Ohlhorst, Craig W., and Sawyer, James W., "Interlaminar Tensile Test Methodology For Two-Directional Carbon-Carbon Composites," <u>NASA CP 2482</u>, January, 1987.

2.   Sawyer, James Wayne, "Investigation Of Test Techniques For Measuring Interlaminar Shear Strength Of Two-Dimensional Carbon-Carbon Composites," <u>NASA TM 100647</u>, September, 1988.

3.   Sawyer, James Wayne, "Analytical and Experimental Results on Coated and Uncoated Stiffened Carbon-Carbon Compression Panels," presented at the Metal Matrix, Carbon, and Ceramic Matrix Composites Conference, Cocoa Beach, Florida, January 18-21, 1989.

4.   <u>Compression Properties Of Rigid Plastics</u>, ASTM D695-80.

5.   While, Don M., "Advanced Carbon-Carbon (ACC) Coating Improvements Summary," <u>NASA CR-172271</u>, December, 1983.

6.   Kelly, A., <u>Strong Solids</u>, Appendix A, 2nd edition, Clarendon Press, Oxford, 1973.

TABLE I. SUMMARY OF DIMENSIONS AND COMPRESSION TEST RESULTS FOR UNCOATED 0/90° LAMINATE ACC TYPE CARBON-CARBON SPECIMENS.

Specimen no.	Specimen configuration	Potted ends	Length, inches	Width, inches	No. of plies	Thickness, inches	Failure stress, ksi	Modulus, Msi
1	A	yes	3.05	0.754	9	0.108	24.0	15.7
2	A	yes	3.05	0.751	9	0.108	23.8	15.7
3	A	yes	2.55	0.750	9	0.107	25.4	16.6
4	A	yes	2.55	0.751	9	0.107	24.8	17.4
5	A	no	2.55	0.749	9	0.106	24.3	16.8
6	A	no	2.55	0.750	9	0.106	23.4	16.7
7	A	no	2.55	0.750	9	0.107	23.8	15.3
8	A	yes	2.55	0.500	9	0.102	25.1	16.6
9	A	yes	2.55	0.500	9	0.103	26.1	16.7
10	A	yes	2.55	0.500	9	0.103	25.6	16.6
11	A	no	2.55	0.501	9	0.104	24.5	16.9
12	A	no	2.55	0.501	9	0.104	24.6	17.0
13	A	no	2.55	0.500	9	0.104	23.9	17.4
14	A	no	2.55	0.252	9	0.105	24.9	20.8
15	A	no	2.55	0.253	9	0.105	25.6	19.6
16	A	no	2.55	0.251	9	0.107	24.0	14.9
17	B	yes	2.55	0.751	9	0.109	25.6	15.8
18	B	yes	2.55	0.752	9	0.104	26.3	16.5
19	B	yes	2.55	0.753	9	0.105	25.1	17.7
20	B	no	2.55	0.752	9	0.105	24.7	16.9
21	B	no	2.55	0.752	9	0.105	24.7	16.3
22	B	no	2.55	0.752	9	0.105	24.5	17.8
23	B	yes	2.55	0.501	9	0.104	24.4	16.7
24	B	yes	2.55	0.500	9	0.105	24.9	16.6
25	B	yes	2.55	0.501	9	0.104	25.6	16.6
26	B	no	2.55	0.500	9	0.104	24.7	17.5
27	B	no	2.55	0.501	9	0.104	24.2	16.7
28	B	no	2.55	0.502	9	0.105	23.5	16.7

TABLE II. SUMMARY OF DIMENSIONS AND COMPRESSION TEST RESULTS FOR UNCOATED ±45° LAMINATE ACC TYPE CARBON-CARBON SPECIMENS.

Specimen no.	Specimen configuration	Potted ends	Length, inches	Width, inches	No. of plies	Thickness, inches	Failure stress, ksi	Modulus, Msi
1	A	no	2.55	0.750	9	0.103	8.1	3.1
2	A	no	2.55	0.751	9	0.104	8.2	3.3
3	A	no	2.55	0.751	9	0.105	8.3	3.4
4	A	no	2.55	0.500	9	0.104	8.1	3.1
5	A	no	2.55	0.500	9	0.105	8.1	2.9
6	A	no	2.55	0.501	9	0.105	8.3	2.8
7	A	no	2.55	0.252	9	0.109	8.2	3.3
8	A	no	2.55	0.254	9	0.109	8.0	3.3
9	A	no	2.55	0.252	9	0.109	7.5	3.4

A - dog bone shape with the base 0.25 inches wider than the gauge section and with a 0.75 inch shoulder radius
B - rectangular shape

TABLE III. SUMMARY OF DIMENSIONS AND COMPRESSION TEST RESULTS FOR COATED 0/90° LAMINATE ACC CARBON-CARBON SPECIMENS.

Specimen no.	Specimen configuration	Potted ends	Length, inches	Width, inches	No. of plies	Thickness, inches	Failure stress, ksi	Modulus, Msi
1	A	yes	3.05	0.75	6	0.074	24.3	15.1
2	A	yes	3.05	0.75	6	0.074	24.1	15.5
3	A	yes	3.05	0.75	6	0.074	24.6	14.4
4	A	yes	3.05	0.5	6	0.074	24.6	15.5
5	A	yes	3.05	0.5	6	0.074	22.4	14.8
6	A	yes	3.05	0.5	6	0.074	24.2	14.0
7	A	yes	3.05	0.25	6	0.074	24.5	15.5
8	A	yes	3.05	0.75	12	0.137	25.2	16.6
9	A	yes	3.05	0.75	12	0.136	25.0	16.8
10	A	yes	3.05	0.75	12	0.137	27.4	16.7
11	A	yes	3.05	0.5	12	0.139	26.9	16.1
12	A	yes	3.05	0.5	12	0.136	28.5	16.8
13	A	yes	3.05	0.25	12	0.137	26.9	16.1
14	A	yes	3.05	0.25	12	0.137	27.1	15.6
15	A	yes	3.05	0.25	12	0.138	26.5	16.2
16	A	yes	3.05	0.25	12	0.136	26.6	15.7

TABLE IV. SUMMARY OF DIMENSIONS AND COMPRESSION TEST RESULTS FOR COATED ±45° LAMINATE ACC CARBON-CARBON SPECIMENS.

Specimen no.	Specimen configuration	Potted ends	Length, inches	Width, inches	No. of plies	Thickness, inches	Failure stress, ksi	Modulus, Msi
1	A	yes	3.05	0.75	6	0.075	15.5	5.4
2	A	yes	3.05	0.75	6	0.075	15.3	5.5
3	A	yes	3.05	0.75	6	0.075	15.2	5.5
4	A	yes	3.05	0.5	6	0.074	15.6	5.4
5	A	yes	3.05	0.5	6	0.076	16.3	4.7
6	A	yes	3.05	0.5	6	0.074	15.0	5.1
7	A	yes	3.05	0.25	6	0.075	13.2	5.1
8	A	yes	3.05	0.25	6	0.075	12.2	4.9
9	A	yes	3.05	0.25	6	0.075	14.9	5.3
10	A	yes	3.05	0.75	12	0.135	14.4	5.4
11	A	yes	3.05	0.75	12	0.137	14.8	6.1
12	A	yes	3.05	0.75	12	0.136	14.0	5.4
13	A	yes	3.05	0.5	12	0.136	14.1	6.1
14	A	yes	3.05	0.5	12	0.136	15.0	5.9
15	A	yes	3.05	0.5	12	0.136	14.0	6.2
16	A	yes	3.05	0.25	12	0.137	13.7	5.9
17	A	yes	3.05	0.25	12	0.137	13.0	5.5

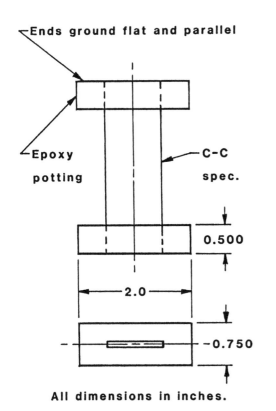

Figure 1. Potted-end compression
specimen configuration.

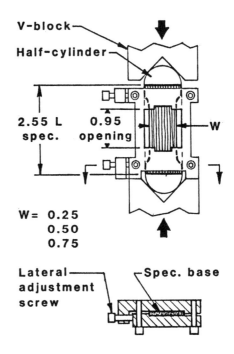

Figure 2. NASA-Langley test fixture
for non-potted specimens.

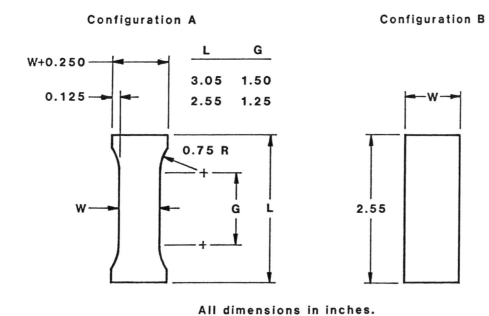

Figure 3. Compression specimen configurations.

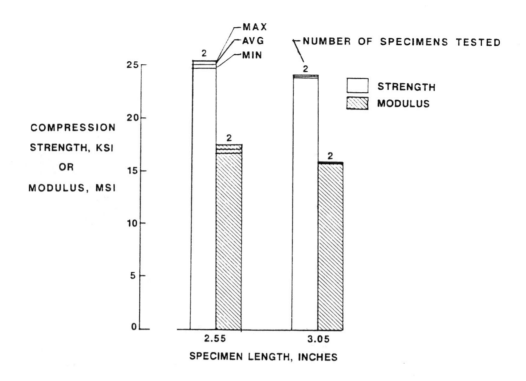

Figure 4. Compression strength and modulus for uncoated 0/90° cross-ply carbon-carbon composite material. Specimen configuration A with potted ends and width of 0.750 inches.

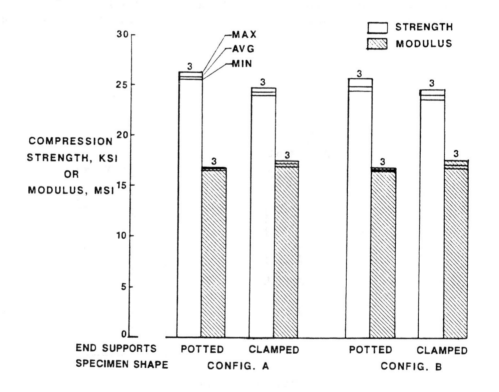

Figure 5. Compression strength and modulus for uncoated 0/90° cross-ply carbon-carbon composite material. Specimens are 2.55 inches long by 0.500 inches wide.

240

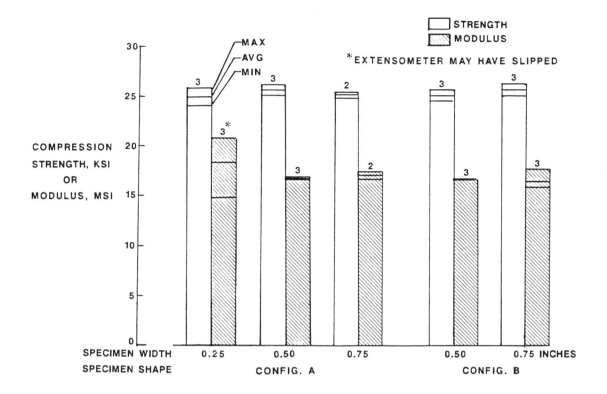

Figure 6. Compression strength and modulus for uncoated 0/90° cross-ply carbon-carbon composite material. Specimens have potted ends and length of 2.55 inches.

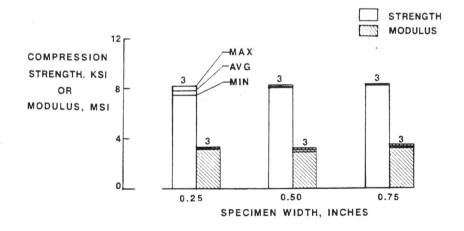

Figure 7. Compression strength and modulus for uncoated ±45° cross-ply carbon-carbon composite material. Specimen configuration A with potted ends and length of 2.55 inches.

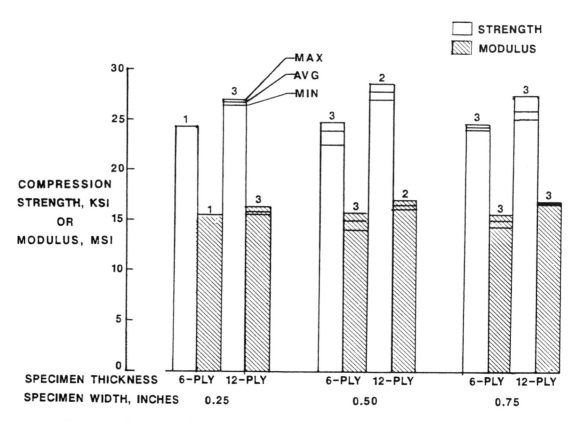

Figure 8. Compression strength and modulus for coated 0/90° cross-ply carbon-carbon composite material. Specimen configuration A with potted ends and length of 3.05 inches.

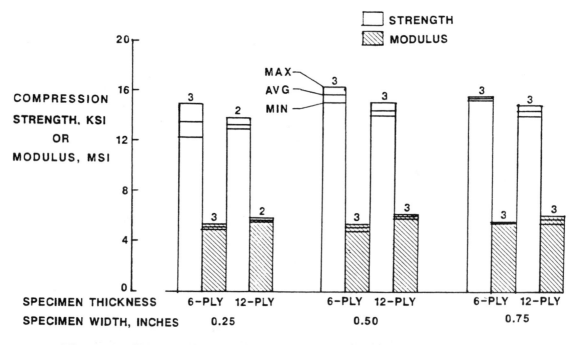

Figure 9. Compression strength and modulus for coated ±45° cross-ply carbon-carbon composite material. Specimen configuration A with potted ends and length of 3.05 inches.

Effects of Crimp Angle on the Tensile Strength of a Carbon-Carbon Laminate

JULIUS JORTNER*

ABSTRACT

For one carbon-carbon laminate, reinforced with plain-woven cloth, this paper shows that variations in the tensile strength are due largely to variations in the crimp angle of the interwoven yarns of the cloth reinforcement. The crimp angle is defined as the maximum angular deviation between the local yarn direction and the plane of the cloth layer. When crimp is characterized by measuring crimp angles for each layer of cloth in a test specimen, it is found that low strengths are associated with large values of average crimp angle; small values of average crimp angle are associated with high strengths. The observed trend of tensile strength versus average crimp angle is fairly well fit with an equation that states strength is inversely proportional to the sine of the average crimp angle; this suggests that the dominant cause of fracture is shear acting on planes containing the crimp direction.

These findings imply that crimp angles must be known within a couple of degrees if accurate strength predictions are to be possible for such laminates. For critical applications, it seems necessary to control crimp angles during manufacture and to be able to measure or verify crimp angles nondestructively in structural parts.

Further studies, including efforts to correlate other properties to crimp, and theoretical attempts to understand crimp-related failure modes, appear warranted.

INTRODUCTION

The material studied here is reinforced with a plain-woven fabric of low-modulus carbon yarns derived from rayon. The cloth layers in the laminate are "warp-aligned"; that is, they are stacked so the warp yarns of each layer are parallel to warp yarns in the neighboring layers. The yarns running perpendicular to the warps are known as fill yarns. Fill and warp yarns have the same number of fibers per yarn; however, the weave is unbalanced, having about 22 fill yarns and about 28 warp yarns per inch of cloth. The cloth is prepregged with a phenolic-base resin before being laminated. Figure 1 shows photomicrographs of the prepreg.

* Jortner Research & Engineering, Inc, Box 2825, Costa Mesa, CA 92628

The carbon matrix derives from the phenolic resin used to prepreg the cloth, and from a liquid impregnant, perhaps a mixture of resin and pitch. The laminates discussed in this paper have all received a final heat treatment to about 2500 C. The bulk density of the finished laminate is about 1.50 g/cc. Figure 2 shows photomicrographs of the composite. Typically, there are about 80 cloth layers per inch of laminate thickness.

Tensile strengths at room temperature, measured on several batches of this material, are summarized in Fig. 3. The variability of strengths in the fill and warp directions of the laminate is large; the higher values are more than 50 percent greater than the lower values, in each direction.

In a preliminary attempt to find the cause of the observed strength variability, several tensile specimens were examined with a scanning electron microscope (SEM). Figure 4 shows SEM views of two specimens tested in the fill direction. It is clear from these photos that the fill yarns are "wavier" in the low-strength specimen than in the high-strength specimen. This observation led to the following study of crimp angles.

MEASUREMENT OF MICROSTRUCTURAL FEATURES

Yarns in a cloth are crimped because they pass over and under each other. In a plain weave, the waveform of the crimped yarns is similar to a sine wave with a wavelength covering two of the yarns being crossed. If the weave were perfectly regular, the maximum angle of crimp (Figure 5) would be the same for all yarns in a given direction (warp or fill). In

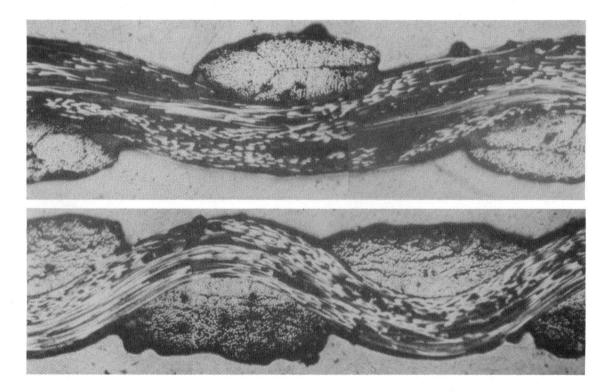

Fig. 1. Photomicrographs of polished sections of prepregged cloth. Top view shows crimp in a warp yarn; bottom view shows crimp in a fill yarn. Note the fiber bundles are twisted and appear to comprise three yarn plies (see fill-yarn cross-sections in top photo). Porous zones at edges of the prepreg are due to interaction between the (uncured) prepreg resin and the mounting resin. Magnification in both photos is the same.

actuality, as in Figures 2 and 4, variations in crimp occur along a yarn, from layer to layer of cloth, and from specimen to specimen.

Theoretically, stiffness and strength of the composite, measured in the direction of one set of yarns, are influenced by the volume fraction of fibers in the yarns oriented parallel to the applied stress and by the waviness of those fibers [1,2]. As all the yarns in Material A are said to contain the same number of fibers, volume fraction is represented here by the number of load-oriented yarns per unit cross-sectional area. Waviness is represented by an average crimp angle for the laminate.

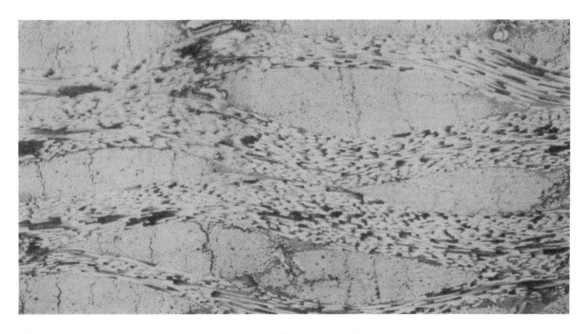

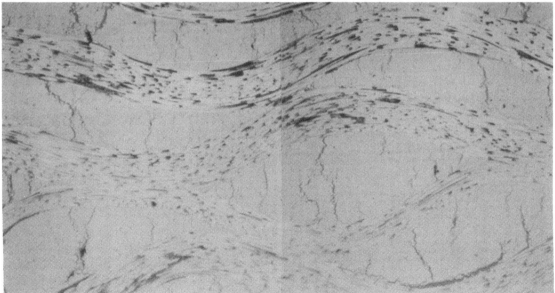

Fig. 2. Photomicrographs of the carbon-carbon laminate. Top view shows the waviness of warp yarns in the warp-normal plane; bottom view shows waviness of fill yarns in the fill-normal plane. Magnification is same as in Fig. 1. Note yarn cross-sections have characteristic microcracks oriented approximately normal to the yarn directions.

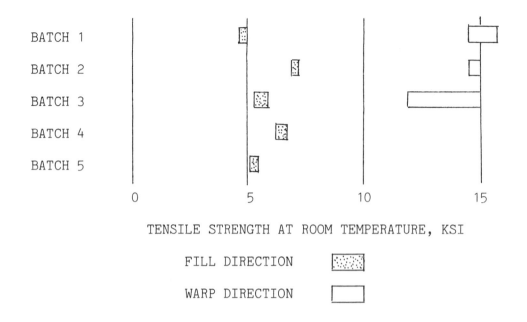

TENSILE STRENGTH AT ROOM TEMPERATURE, KSI

FILL DIRECTION

WARP DIRECTION

Fig. 3. Tensile strengths of various batches of Material A.

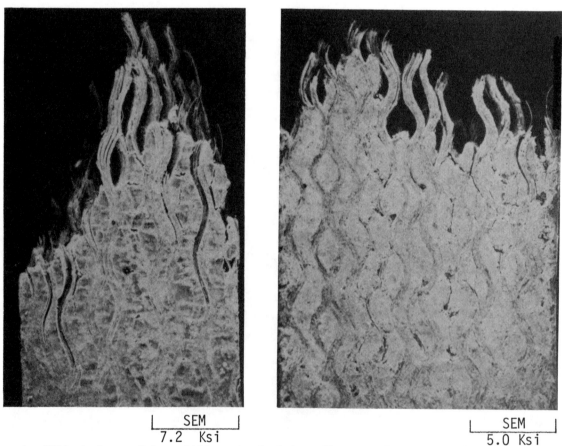

SEM
7.2 Ksi

SEM
5.0 Ksi

Fig. 4. SEM photos of two fill-oriented tensile specimens showing vicinity of fracture. Failure stress is noted for each specimen. The stronger (left) shows less crimp in the fill yarns than the weaker specimen (right).

246

The procedure for estimating these microstructural quantities (defined in Fig. 6) is:

a. Count the number of cloth layers through the depth of the gage section of the specimen.

b. Count the number of wavelengths of transverse yarns across the specimen width and multiply by 2 to estimate the number of load-bearing yarns in one cloth layer.

c. Estimate the number of load-bearing yarns per square inch of cross-section: multiply the number of yarns (from b) by the number of layers (from a) and divide by the product of depth and width.

d. Estimate the average maximum crimp angle: On one region of a side surface, polished to about the 3-micron grit level and observed under about 75X magnification, measure one positive and one negative maximum crimp angle on each yarn visible on the surface. Thus, the number of measurements is close to twice the number of layers. The absolute values of these angles were averaged to produce the value called the average maximum crimp angle. Each measurement was done by rotating the microscope's ocular lens to align a crosshair with the local yarn direction; the angle was estimated with the aid of a protractor scale attached to the ocular. Zero angle was aligned with the specimen edge (for coupons) or with the mean yarn direction (estimated by eye) for samples from involute rings; the precise zero alignment is not crucial because the positive and negative angles are averaged before use in correlations. From repeated measurements on the same sample, the repeatability of individual measurements seems within about two degrees, and the average angle seems repeatable within about plus or minus one degree. However, measurements on a different region of the same sample would produce slightly different numbers. In the reported measurements,

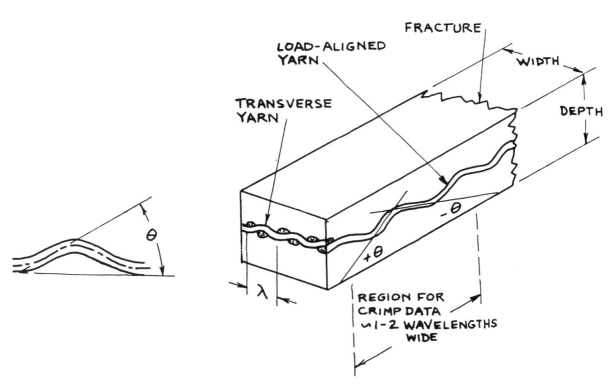

Fig. 5. Definition of maximum crimp angle, θ.

Fig. 6. Nomenclature for measured features.

the ocular crosshair was aligned parallel to the centerline of the visible portion of the yarn bundle; hence, alignment was with the boundaries of the yarn rather than with the filaments (which may be at different angles because the fiber bundles are twisted). Judgment is involved in discerning the maximum crimp angle so another investigator might see slightly different angles, especially for the warp yarns, which often are wiggly on a sub-wavelength scale (Fig. 2).

Aside from questions of technique and repeatability, there should be concern whether the average maximum crimp angle is the best parameter for correlating strengths. It was used here because it is relatively easy to quantify. Theoretically [1,2], a better approach might be to characterize the entire waveforms of the crimped yarns and derive a number that would reflect the effective stiffness of the composite.

CORRELATION OF TENSILE STRENGTHS WITH CRIMP

Tensile specimens and strength data were obtained from several investigators (see Acknowledgments). The specimens include coupons from flat panels tested under uniaxial tensile load, and involute [3] rings tested under internal pressure. These were examined for crimp and number of load-bearing yarns per unit cross- sectional area. The results, including reported tensile strengths, are listed in Table I.

Figure 7 shows the trend of ultimate load per yarn versus average maximum crimp angle. Warp- and fill-oriented data lie on the same trend; all the data is reasonably well fit by a curve whose equation is:

$$L = K/\sin\theta \qquad\qquad (1)$$

where L is the ultimate load per yarn, θ is the average maximum crimp angle, and K is a constant estimated as 1.55 pounds for Material A. Tensile stress at failure, σ , may be estimated as

$$\sigma = LN \qquad\qquad (2)$$

where N is the number of yarns per unit cross-sectional area.

A physical interpretation of Eq. 1 is that failure initiates by shear on planes containing the crimp angle (Fig. 8) for yarns having a shear capacity proportional to the cosine of the crimp angle. That is, τ_θ , the shear stress acting on those planes is $\sigma \sin\theta \cos\theta$, which in combination with equations 1 and 2 may be manipulated to show that Equation 1 implies $\tau_\theta/N\cos\theta$ is a constant at tensile failure. Although suggestive, this observation is not yet an explanation; a mechanistic explanation for Eq. 1 is being attempted and may be reported in due course.

Equation 1 is not a perfect fit. In particular, warp tension strengths at lower crimp angles are underestimated. Perhaps correlations should be developed separately for warp and fill yarns. However, as a first-cut working correlation, Equation 1 seems capable of predicting tensile strengths of Material A within about 15 percent.

IMPLICATIONS AND CONCLUDING REMARKS

Average maximum crimp angles in Material A vary from about 20 to 30 degrees in the fill yarns, and from about 12 to 18 degrees in the warp yarns. This variability of crimp angle appears responsible for much of the variation in strengths that is observed in this material. Fig. 1

shows a good correlation between crimp angles and tensile strengths measured in the fill and warp directions. Other properties of the laminate might also depend on crimp, but the required correlations have yet to be obtained.

Variability of yarn count (number of yarns in each direction per unit cross-sectional area) has been implicitly included in the strength-crimp correlations, by expressing strength in terms of load per yarn (Table I). The observed variability of yarn count contributes to the variability of stress at failure, but to a lesser extent than crimp variability.

One implication of these findings is that stress analyses of structural parts made of Material A would be more accurate if the input properties were adjusted for each part being analyzed. In other words, because of the large variability of crimp and crimp-related properties, either the crimp angles or the properties should be measured for each individual part to provide reasonably accurate inputs to structural analyses.

Table I. Microscope Measurements of Laminate Weave Parameters in Tensile Specimens
         - Correlation to Tensile Strengths

Spec Number	Layer Count	Depth in.	Layers per inch	Yarn No.	Width in.	Yarns per in.	Yarn Count per in	Avg Crimp Angle deg	Reported Strength psi	Avg Yarn Load lb	Specimen Source
FILL TESTS											
1-1F	20.0	0.251	80	15.6	0.747	20.9	1660	31	4979	3.00	Batch 1, Coupon
1-2F	18.0	0.250	72	14.0	0.673	20.8	1500	28	4745	3.16	Batch 1, Coupon
1-3F	18.5	0.251	74	14.0	0.685	20.4	1510	29	4788	3.17	Batch 1, Coupon
2-1F	12.0	0.171	70	11.0	0.500	22.0	1540	20	7210	4.68	Batch 2, Coupon
3-1F	16.5	0.248	67	8.0	0.396	20.2	1340	26	5333	3.98	Batch 3, Coupon
3-2F	19.0	0.252	75	16.0	0.755	21.2	1600	25	5653	3.53	Batch 3, Coupon
3-3F	18.0	0.257	70	15.0	0.737	20.4	1430	22	5856	4.10	Batch 3, Coupon
4-1R	38.0	0.502	76	10.8	0.498	21.7	1640	23	6825	4.16	Batch 4, Involute Ring
4-2R	37.5	0.503	75	10.6	0.499	21.2	1580	25	6462	4.09	Batch 4, Involute Ring
4-3R	39.0	0.503	78	10.7	0.495	21.6	1680	27	5168	3.08	Batch 5, Involute Ring
4-4R	37.5	0.503	75	10.5	0.500	21.0	1570	29	5178	3.30	Batch 5, Involute Ring
WARP TESTS											
1-1W	19.0	0.251	76	18.6	0.674	27.6	2090	15	15652	7.49	Batch 1, Coupon
1-2W	19.0	0.251	76	19.0	0.666	28.5	2160	14	15564	7.21	Batch 1, Coupon
1-3W	20.0	0.251	80	19.2	0.680	28.2	2250	14	14517	6.45	Batch 1, Coupon
2-1W	12.0	0.171	70	13.4	0.500	26.8	1880	13	14970	7.96	Batch 2, Coupon
3-1W	19.0	0.252	75	21.0	0.747	28.1	2120	18	10720	5.06	Batch 3, Coupon
3-2W	19.0	0.255	75	21.6	0.765	28.2	2100	17	10015	4.77	Batch 3, Coupon
3-3W	19.5	0.252	77	21.0	0.753	27.9	2160	16	11650	5.39	Batch 3, Coupon

Notes:

a. Layer count refers to number of cloth layers through depth of specimen, counted at one surface.

b. Yarn no. refers to number of yarns in load direction across width of specimen, in one cloth layer; to improve accuracy, the count was usually made at a tab end, which is wider than the gage section.

c. Yarn count, the number of yarns per square inch is the product of layers per inch and yarns per inch; this is an estimate of the number of loaded yarns per inch of crosssection.

d. Average crimp angle is the average of positive and negative maximum crimp angles measured from one traverse across the depth of each specimen. Angle measurements refer to the yarn bundle angles, not filament angles. Absolute value of negative angles was used in average.

e. Yarn load is the average load on each load-oriented yarn at fracture, calculated as the quotient of strength and number of yarns per square inch.

f. In tests of involute rings, because of the arc angle of the involute layup [3], test direction is about 8 degrees off the fill yarns; Stress listed is max. fill stress calculated from stress state at inner radius of ring, based on elastic (isotropic Lame) analysis of internal pressure at fracture.

The causes of crimp variability are not understood yet. We do not know at what stage of composite fabrication the crimp angles are set. Is the variability caused during weaving, during heat-treatment of the fabric, during the prepregging operation, or during laminate debulking and cure? The available microstructural measurements reveal no consistent simple relationships between warp crimp angle and fill crimp angle, or between crimp angles and yarn count or ply spacing. It seems worthwhile to investigate the matter, for the sake of developing ways of reducing the crimp variability.

If crimp is found to vary significantly within individual parts, it would seem necessary to develop nondestructive means for estimating in-situ crimp angles within critical components.

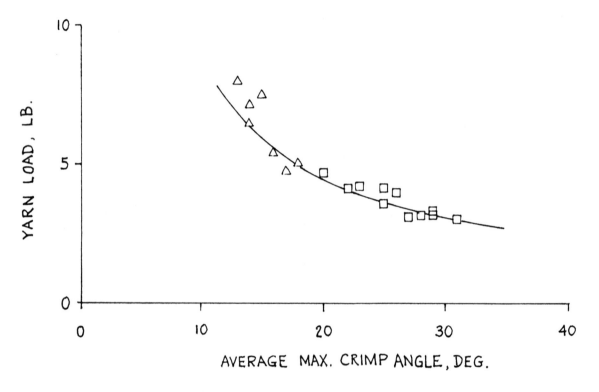

Fig. 7. Tensile load capacity of yarns as a function of average maximum crimp angle. Material A, data from Table I. Curve is Equation 1. Triangles are warp data; squares are fill data.

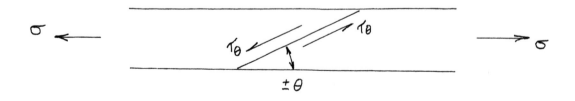

Fig. 8. Definition of T_θ .

Better theoretical understanding of the crimp effect on strength would be desirable. Study of crimp effects in other carbon-carbon laminates, including those reinforced with satin and twill weaves, would be worthwhile. To support such work, the development of better ways of measuring and characterizing crimp, perhaps using computer-aided image analysis methods, would be welcome.

ACKNOWLEDGMENTS

This work was sponsored by the Air Force Astronautics Laboratory at Edwards Air Force Base. It was initiated under a subcontract to Aerojet Strategic Propulsion Co under Air Force contract F04611-85-C-0099. Additional support came from Boeing Aerospace Co and from PDA Engineering under Air Force contracts F04611-86-C-0080 and F04611-86-C-0050, respectively. Tensile specimens were supplied by D. Bocek and J. Kirkhart of Aerojet, E. Stokes and B. Wingard of the Southern Research Institute, and D. Marx of PDA Engineering. The carbon-carbon laminates were manufactured by Kaiser Aerotech, San Leandro, California. The programs were managed by S. Durham of Aerojet, J. Nelson of Boeing, and J. G. Crose of PDA, and monitored by J. Hildreth and T. Galati of the Air Force Astronautics Laboratory. The author thanks these individuals and organizations for their cooperation and support.

REFERENCES

1. Jortner, J., "A Model for Predicting Thermal and Elastic Constants of Wrinkled Regions in Composite Materials," Effects of Defects in Composite Materials, ASTM STP 836, 1984, pp. 217-236.

2. Jortner, J., "A Model for Nonlinear Stress-Strain Behavior of 2D Composites with Brittle Matrices and Wavy Yarns," Advances in Composite Materials and Structures (S. S. Wang and Y. D. S. Rajapakse, Eds.), ASME AMD-82, 1986, pp. 135-146.

3. Pagano, N. J., "General Relations for Exact and Inexact Involute Bodies of Revolution," Advances in Aerospace Structures and Materials (R. M. Laurenson and U. Yuceoglu, Eds.), ASME AD-03, 1982, pp. 129-137.

A Criterion for Delamination of Cloth-Reinforced Brittle-Matrix Laminates under Multiaxial Stresses

JULIUS JORTNER,* J. G. CROSE,** D. A. MARX** AND H. S. STARRETT***

ABSTRACT

A criterion for delamination of orthotropic laminates is proposed in terms of the three stress components acting on the interlaminar plane. The criterion, called "WUFR", is a generalization to three dimensions of Wu's criteria (Wu, 1963 and 1965). Strength data for two cloth-reinforced carbon-matrix laminates tested under combinations of interlaminar shear and crossply tension or compression are well represented by the criterion Several unresolved issues and needs for further work are discussed.

INTRODUCTION

Delamination is a dangerous mode of failure for brittle-matrix laminates because the fracture proceeds along the unreinforced layer of matrix between the fibrous layers. A criterion for delamination is needed to support analytical predictions of the structural capabilities of components made with such laminates. Although the work described below deals with carbon-matrix laminates, the results may prove applicable also to other brittle-matrix composites.

The composites considered here are laminates of carbon cloth in a carbon matrix. The cloth layers contain mutually perpendicular sets of fiber bundles referred to as warp and fill yarns. Each layer is oriented similarly to every other layer, in what is referred to as a warp-aligned laminate. Because they are warp-aligned, these laminates do not seem to exhibit the edge-delamination problem characteristic of angle-ply laminates (eg,[1]). Thus, delamination is viewed here as being caused directly by interlaminar stress components of the laminate stress state without contribution from minimechanical ply interactions.

To estimate interlaminar margins of safety, analysts frequently use the quadratic equation of Tsai and Wu [2], in a form that assumes in-plane stress components have no effect:

$$F_3 \sigma_{33} + F_{33} \sigma_{33}^2 + F_{44} \sigma_{23}^2 + F_{55} \sigma_{13}^2 = 1 \qquad (1)$$

* Jortner Research & Engineering, Inc, Box 2825, Costa Mesa, CA 92628
** PDA Engineering, 2975 Redhill Ave, Costa Mesa, CA 92626
*** Southern Research Institute, Box 55305, Birmingham, AL 35255

where σ are stress components and the F factors are constants usually quantified in terms of the relevant uniaxial strengths. The subscripts 1, 2, and 3 refer respectively to the warp, fill, and normal (also called crossply) directions. This interlaminar Tsai-Wu criterion is an ellipsoid in stress space. Although the quadratic formula produces a reasonable fit to most available data, it does not have an explicit mechanistic basis and does not promise that delamination will be the failure mode for all points on the ellipsoidal surface. For example, under high levels of crossply compression, we would expect delamination to be suppressed and failure to occur by some other mechanism, perhaps translaminar shear or crushing.

THE "WUFR" CRITERION

An alternate formulation for delamination failures is proposed here. We assume delamination initiates at a crack-like flaw between laminae of the 2D composite. The work of Wu [3], summarized by Corten [4], suggests the following criterion, when the crossply stress is positive (tensile):

$$\left(\frac{K_I}{K_{Ic}}\right)_{33} + \left(\frac{K_{II}}{K_{IIc}}\right)^2_{44} + \left(\frac{K_{II}}{K_{IIc}}\right)^2_{55} = 1 \qquad (2)$$

where the K's are mode I and mode II stress intensity factors and the K_c's are the critical values for crack propagation under the normal and shear stress components indicated by subscripts. Assuming the flaws are inherent in the material, and that they do not grow under loading before fracture, Eq. 2 may be expressed directly in terms of applied stresses:

$$\frac{\sigma_{33}}{T_{33}} + \left(\frac{\sigma_{23}}{S_{23}}\right)^2 + \left(\frac{\sigma_{13}}{S_{13}}\right)^2 = 1 \qquad (3)$$

where T and S respectively denote tensile and shear strengths.

Wu's criterion is shown by Hahn [5] to be a special case of a broader mixed-mode formulation. We adopt Wu's criterion because Hahn shows it fits (the rather scarce) available data for fiber-reinforced laminates quite well (also see Corten [4]).

When crossply stress is compressive, the presence of K_I in Eq. 2 is unwarranted inasmuch as mode I propagation would not occur. In this case, friction induced by the compressive stress across the flaw would allow some transfer of shear stress across the flaw, as described by McClintock and Walsh [6], Wu [7] and Corten [4]. The criterion under crossply compression then simplifies to

$$\sqrt{\sigma^2_{23} + \sigma^2_{13}} + \mu\sigma_{33} = \tau' \qquad (4)$$

where τ' is the magnitude of the stress vector that touches the failure surface and has the same direction as the projection of the applied stress vector onto the plane of zero crossply stress. The parameter μ is the effective friction coefficient. τ' and μ are obtained from

$$\tau' = S_{13}\sqrt{G} \qquad (5)$$

and

$$\mu = \mu_w\sqrt{H} \qquad (6)$$

where μ_w is the friction coefficient when $\sigma_{23} = 0$, and

$$G = \frac{1 + \left|\frac{\sigma_{13}}{\sigma_{23}}\right|^2}{\left|\frac{S_{13}}{S_{23}}\right|^2 + \left|\frac{\sigma_{13}}{\sigma_{23}}\right|^2} \qquad\qquad H = \frac{1 + \left|\frac{\sigma_{13}}{\sigma_{23}}\right|^2}{\left|\frac{\mu_w}{\mu_F}\right|^2 + \left|\frac{\sigma_{13}}{\sigma_{23}}\right|^2}$$

where μ_F is the friction coefficient when $\sigma_{13} = 0$. Both G and H are 1 when $\sigma_{23} = 0$. Equation 6 allows for different frictions in the warp and fill directions, and maintains an elliptical cross-section of the failure surface at any constant crossply stress.

The criterion represented by equations 3 and 4, is referred to as the Wu mixed-mode plus friction model, or WUFR for short. For tensile crossply stresses, WUFR is an ellipsoidal failure surface in stress space. Thus it may be viewed as a special case of the Tsai-Wu quadratic (Eq. 1) in this region of stress space; that is, Eq. 3 may be derived from Eq. 1 by taking the crossply compressive strength to be infinite. For compressive crossply stresses, the criterion has an elliptic cross-section at constant crossply stress, but the variation of interlaminar shear stress at failure with crossply stress is linear because of the assumption of Newtonian frictional effects. Figure 1 is a sketch of the WUFR failure surface.

The WUFR criterion may be quantified using five material properties: tensile strength in uniaxial crossply tension, shear strength in fill-normal shear, shear strength in warp-normal shear, coefficient of friction for fill-normal shear, and coefficient of friction for warp-normal shear. Of these, crossply tensile strength is most readily determined, by testing the material in crossply tension. The remaining properties are less directly determined because of difficulties in testing for pure uniaxial shear. That is, real shear tests always seem to involve some degree of crossply stress. Assuming that the shear tests can be analyzed

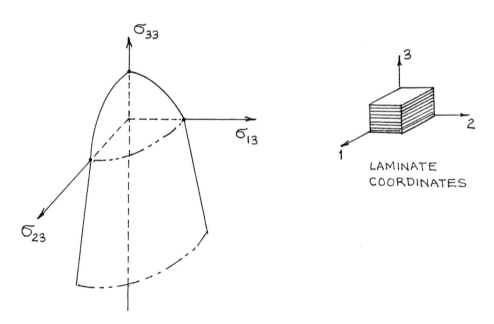

Fig. 1. Schematic of the WUFR Criterion for Delamination.

to estimate the stress state at failure, the required properties can be estimated from tests in five stress states. For example, the five stress states might be:

1) fill-normal shear with a small degree of crossply compression;
2) fill-normal shear with a larger degree of crossply compression;
3) warp-normal shear with a small degree of crossply compression;
4) warp-normal shear with a larger degree of crossply compression; and
5) crossply tension.

Fitting a straight line to a graph of shear strengths vs crossply stress measured in the fill-normal tests would give μ_F as the slope of the line and S_{23} as its intercept at zero crossply stress. The same procedure would apply to the warp-normal tests to give μ_W and S_{13}.

CORRELATION OF THE WUFR CRITERION WITH TEST DATA

Delamination strengths were measured for two carbon-carbon materials. Material A is a flat laminate of plain-weave (2-harness) cloth made with low-modulus carbon fibers derived from rayon, densified to about 1.5 g/cc. The cloth is somewhat unbalanced; there are about 22 fill yarns per inch of cloth, compared to about 28 warp yarns per inch. Material B is a flat laminate of 8-harness satin-weave cloth made with intermediate-modulus carbon fibers derived from poly-acrylonitrile, densified to approximately 1.65 g/cc. The cloth has about 23 fill yarns per inch and about 24 warps per inch. Both composites exhibit nonlinear stress-strain responses, some of which have been described by Stanton and Kipp [8] and Jortner [9].

Tests of several specimen-and-loading configurations provided delamination strength data under various combinations of interlaminar shear and crossply loadings. Crossply tensile strength was obtained by simple tensile tests of cylindrical coupons. Tests in combined loadings used a modified direct shear method and a tube-torsion method.

The modified direct shear (MDS) test (Fig. 2) provides a fairly large region of almost uniform stresses in the failure region (Fig. 3). Stresses at failure are estimated from the loads at failure using equations derived from a series of finite-element analyses:

$$\text{Material A, fill-normal tests:} \qquad \tau = \bar{\tau} (1.12 - .12R) \qquad (7)$$

$$\text{Material A, warp-normal tests:} \qquad \tau = \bar{\tau} (1.10 - .10R) \qquad (8)$$

$$\text{Material B, warp-normal tests:} \qquad \tau = \bar{\tau} (1.03 - .03R) \qquad (9)$$

$$\text{Materials A and B, all tests:} \qquad S_{33} = \tau (.18 - 1.10R) \qquad (10)$$

where τ is maximum shear stress at failure, $\bar{\tau}$ is average (load/area) shear stress at failure, and R is the ratio of lateral to axial axial loads at failure. The nonlinear stress-strain law used in the analyses is that of Stanton and Kipp (1985). Equations 7-10 are approximate fits to the analysis results over a range of load ratios, $0 \leq R \leq .4$; in actuality, the correction factors are stress-level dependent, but the dependency is small; that is, differences between elastic and nonlinear estimates for peak shear and crossply stresses are less than 3 percent in all cases.

Material A also was tested in the form of tubes, cut from blocks, under several combinations of torsion and axial load (Fig. 4). Several

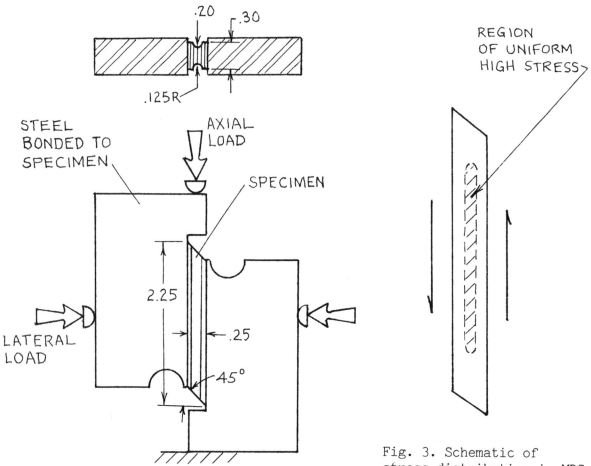

.20
.30
.125R

STEEL BONDED TO SPECIMEN

AXIAL LOAD

SPECIMEN

2.25

.25

LATERAL LOAD

45°

REGION OF UNIFORM HIGH STRESS

Fig. 2. Modified direct shear (MDS) test. Dimensions in inches.

Fig. 3. Schematic of stress-distribution in MDS specimen. Shaded zone has maximum shear and most positive crossply stresses.

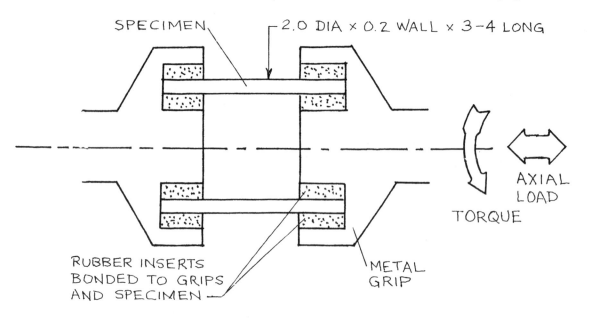

SPECIMEN

2.0 DIA × 0.2 WALL × 3-4 LONG

TORQUE

AXIAL LOAD

RUBBER INSERTS BONDED TO GRIPS AND SPECIMEN

METAL GRIP

Fig. 4. Tube torsion test; schematic omits certain details of grip design that provide for ease of alignment and bonding. Dimensions in inches.

orientations of tubes were used. Across-ply tubes (APT) have the cylindrical axis parallel to the normal (crossply) direction; these provide the bulk of the combined data. Because the fill-normal strength for Material A is significantly lower than the warp-normal shear strength, the APT data is assumed to represent failure in fill-normal shear. Fill tubes (FT) and warp tubes (WT) have the axes parallel to the fill and warp directions, respectively. The FT tests provide strength data for shear in the fill-normal direction; the WT tests give warp-normal shear data. That the FT and WT tubes failed by delamination is because the warp-fill shear strength is substantially higher than the interlaminar shear strengths. Three tubes also were tested with axes parallel to the in-plane direction 45 degrees between the warp and fill directions; these warp-fill tubes (WFT) were tested without axial load and provide a measure of shear strength in the off-axis direction.

Shear stresses at failure in tube tests were estimated from the linear-elastic torsion equation with a correction factor, 1-C; that is, shear stress at failure = (1-C)Tr/J, where T is torque, r is outer radius, and J is the polar moment of inertia. The correction C was derived from finite-difference stress analyses using Ramberg-Osgood representations of stress-strain nonlinearity, which show that nonlinearity of stress-strain responses causes the stresses at the outer radius to be less than would be estimated from the linear-elastic formula. The analytical estimates of C differ for each tube orientation and are stress-dependent, ranging from zero at low stress levels to about 6.5 percent at the highest stresses estimated at specimen failure. As a rule of thumb, C = .04 provides a good estimate of shear stress at failure for most cases. The axial stresses in the tubes were estimated simply as load/area.

Figures 5 to 8 show how the WUFR criterion matches the strength data for Materials A and B. The WUFR input parameters used to provide the trendlines in these Figures are listed in Table I.

Figure 9 shows data from FT specimens tested under combinations of torque and axial compression. The data trend (solid line) suggests there is some influence of in-plane stress on interlaminar strength, and vice-versa; the dotted-line trend would be expected if there were no interaction between in-plane and interlaminar stresses.

Agreement between WUFR and the interlaminar data (Fig. 5-8) is fairly good, considering the scatter in data. Some of the scatter may derive from the use of several billets as source material; perhaps careful evaluation of each specimen as to density, microstructure, yarn crimp angles [10], flaws, etc., would allow better correlation.

Table I. Strength Parameters Used to Quantify the WUFR Criterion

	Material A	B
Normal tensile strength, psi	720	890
Warp-normal shear strength, psi	2400	1680
Fill-normal shear strength, psi	1720	*
Fill-normal friction coefficient	0.53	*
Warp-normal friction coefficient	0.25	0.55

* Fill-normal data was not obtained for Matl B

UNRESOLVED ISSUES

The following paragraphs identify some issues regarding the adequacy of the WUFR criterion that cannot be resolved with the data at hand:

a) The crack-propagation model underlying WUFR implies that the delamination threshold is unaffected by the in-plane stresses. Although the implied decoupling appears reasonable, we should note that experimental substantiation currently does not exist. Models that deal explicitly with the geometric waviness of the delamination surface, a result of the nesting of cloth layers in the laminate (Fig. 10), probably would predict some interactions between in-plane and interlaminar stresses. Fig. 9 suggests some interactions do occur in Material A.

b) In the absence of relevant data, mathematical convenience is the main justification for formulating Eq. 3 and 6 to ensure elliptical cross-

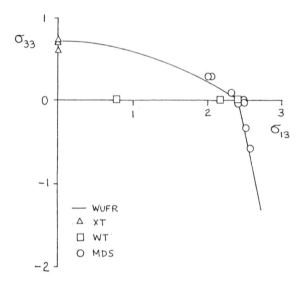

Fig. 5. Warp-normal strengths,
        Material A.

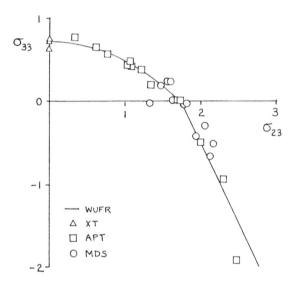

Fig. 6. Fill-normal strengths,
        Material A.

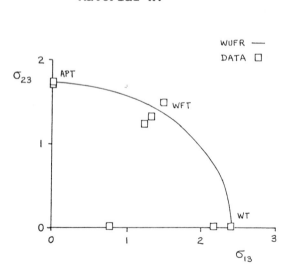

Fig. 7. Interlaminar shear strengths,
        Material A.

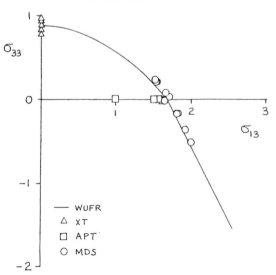

Fig. 8. Warp-normal strengths,
        Material B.

sections of the criterion on planes normal to the crossply coordinate. Very little data has been obtained with regard to interlaminar shear strengths when the shear-stress direction is between the warp and fill directions. Although, Fig. 7 suggests that the WUFR or the Tsai-Wu ellipses are not contradicted by the few data available, more tests are needed to verify the quadratic forms of these criteria, especially in the presence of crossply compression.

c) The large difference between WUFR coefficients for warp-normal and fill-normal frictions in Material A is somewhat unexpected. More warp-normal data would be useful to determine the reality of the apparent frictional difference. Such data would help determine whether WUFR predictions are significantly better than those of the Tsai-Wu criterion (Eq. 1); it is only in the warp-normal data under crossply compression (Fig. 5) that significant discrepancies appear between the data and the Tsai-Wu quadratic, at least over the range of stress-states tested.

d) The specimens used to verify the WUFR criterion were specially designed to eliminate severe stress gradients and to have large regions under fairly uniform high stresses. The applicability of WUFR to delamination of structures that might experience steep stress gradients has not been verified. Although an approach like that of Whitney and Nuismer [11], based on applying the failure criterion to stresses averaged over a characteristic volume of the material, appears reasonable, the applicability of WUFR to structures with steep stress gradients must be considered an open question. Testing of specimens with steep stress gradients appears needed.

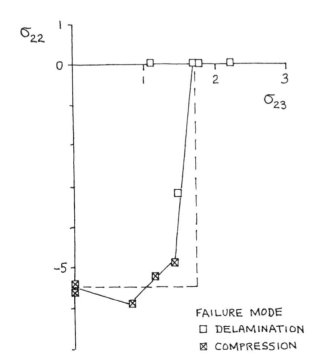

Fig. 9. Strengths of Material A under combinations of fill compression and fill-normal shear, from FT specimens.

Fig. 10. Sketch of Material A showing roughness of delamination surface between cloth plies.

CONCLUDING REMARKS

A delamination criterion called WUFR (for Wu-plus-friction), which is a generalization of equations first proposed by Wu [3,7], is shown to be a reasonable fit to available strength data for two carbon-carbon laminates over a range of ratios of normal stress to shear stress. The WUFR criterion has a mechanistic basis, in contrast to the quadratic form of Tsai and Wu, which it resembles when the normal stress is tensile but differs from when normal stress is compressive. The available data suggest WUFR is a better fit to data than is the Tsai-Wu criterion; however, more data over a larger range of stress combinations would be necessary to verify this tentative conclusion.

Nonlinear stress analyses of test specimens seem essential for derivation of accurate strength information from experiments directed at interlaminar interlaminar shear strengths. This implies that accurate nonlinear constitutive relations must be available for the test material.

Among the unresolved issues identified above, three stand out as requiring further study before the criterion can be considered generally applicable to structural analyses of complex parts: 1) How is delamination resistance influenced by the three in-plane stress components that are currently ignored in the WUFR formulation? 2) How is the criterion to be applied when stress gradients are steep over distances smaller than the repeating unit cell of the material? and 3) How is the observed scatter in strengths to be dealt with in assessing structural safety?

ACKNOWLEDGMENTS

This study was sponsored by the Materials Laboratory of the Air Force Wright Aeronautical Laboratories under contract F33615-85-C-5114 monitored by N. J. Pagano and conducted by PDA Engineering with the assistance of the Southern Research Institute and Jortner Research & Engineering, Inc..

Crossply-tensile and torsion data were obtained by the Southern Research Institute; B. Wingard was the test engineer. J. P. Norman of PDA Engineering conducted the finite-element analyses of the modified direct shear test. The modified direct shear specimens were assembled by R. Hack of PDA and the tests were conducted at Jortner Research & Engineering. The materials were supplied by Kaiser Aerotech, San Leandro, California and Carbon-Carbon Advanced Technologies (C-CAT), Fort Worth, Texas.

The authors thank these individuals and organizations for their contributions and support.

REFERENCES

1. Soni, Som R., and Kim, Ran Y., "Delamination of Composite Laminates Stimulated by Shear," Composite Materials: Testing and Design (7th Conf), ASTM STP 893 (J. M. Whitney, Ed.), ASTM, 1986, pp. 286-307.

2. Tsai, S. W., and Wu, E. M., "A General Theory of Strength for Anisotropic Materials," J. Composite Materials, Vol. 5, January 1971, pp. 58-80.

3. Wu, Edward M., "Application of Fracture Mechanics to Orthotropic Plates", T&AM Report 275, University of Illinois, Urbana, June 1963.

4.   Corten, Herbert T., "Micromechanics and Fracture Behavior of Composites," Modern Composite Materials (L. J. Broutman and R. H. Krock, Eds), Addison-Wesley, 1967, pp.27-93.

5.   Hahn, H. T., "A Mixed-Mode Fracture Criterion for Composite Materials," Comp. Tech. Rev. (ASTM), Vol. 5, No. 1, Spring 1983, pp. 26-29.

6.   McClintock, F. A., and Walsh, J. B., "Friction on Griffith Cracks in Rocks under Pressure," Proc. 4th U.S. Cong. Applied Mechanics, ASME, 1963, pp. 1015-1021.

7.   Wu, Edward M., "A Fracture Criterion for Orthotropic Plates Under the Influence of Compression and Shear," T&AM Report 283, University of Illinois, Urbana, 1965.

8.   Stanton, E. L. and Kipp, T. E., "Nonlinear Mechanics of Two-Dimensional Carbon-Carbon Composite Structures and Materials," AIAA J., Vol. 23, 1985, pp. 1278-1284.

9.   Jortner, J., "A Model for Nonlinear Stress-Strain Behavior of 2D Composites with Brittle Matrices and Wavy Yarns," Advances in Composite Materials and Structures (S. S. Wang and Y. D. S. Rajapakse, Eds.), ASME AMD-82, 1986, pp. 135-146.

10.   Jortner, J., "Effects of Crimp Angle on the Tensile Strength of a Carbon-Carbon Laminate," Symposium on High Temperature Composites, American Society for Composites, June 1989.

11.   Whitney, J. M., and Nuismer, R. J., "Stress Fracture Criteria for Laminated Composites Containing Stress Concentrations," J. Comp. Mat., Vol. 8, July 1974, pp. 253-265.

Characterization of Damage Initiation in Three-Point Bend Specimens of Advanced Woven Carbon-Carbon Composite Materials

P. D. COPP, J. C. DENDIS AND S. MALL

ABSTRACT

Accurate determination of interlaminar shear strength of composite laminated structures has continued to plague designers. The ASTM standard three-point bend test has been routinely employed to measure this critical design data. This study focused on determining damage initiation and failure mechanisms in woven uncoated carbon-carbon composite laminates resulting from the three-point bend test. Failure modes and damage initiation sites for short and long beam geometries were determined. The effects of incremental versus continuous loading were also examined.

INTRODUCTION

The use of advanced materials, such as carbon-carbon composites, for turbine engine applications requires a thorough understanding of their mechanical behavior. It is therefore necessary to have accurate material properties to perform damage and failure analyses. One of the principal properties needed to characterize the delamination behavior of laminated composites is the interlaminar shear strength. Two dimensional woven carbon-carbon composites have a tensile strength approximately forty times greater than their shear strength, thus it is anticipated that shear failures will predominate in design applications.

Interlaminar shear strength is currently estimated by using the ASTM-2344 three-point bend test [1] due to the simplicity of specimen design and testing. Whitney and Browning [2] characterized failures in three- and four- point bend tests of unidirectional graphite epoxy laminates. They concluded that the combined compression and shear stresses present produced localized damage prior to the initiation of any interlaminar shear failure. Whitney [3] substantiated this conclusion with a two dimensional analytical solution using Fourier series to model the applied loads and supports. Copp and Keer [4]

P. D. Copp, Assistant Professor, J. C. Dendis, Former Graduate Student,
S. Mall, Professor, Department of Aeronautics and Astronautics, Air
Force Institute of Technology, Wright-Patterson AFB, OH 45433.

solved the exact mixed boundary value problem of a smooth rigid indenter on a beam of two orthotropic layers supported by point loads using integral transform techniques. Their results showed the stress state was a complex combination of compressive normal, compressive and tensile bending and shear stresses leading to localized damage initiation in the area of the rigid indenter prior to any interlaminar shear failure. Thus, they concluded the ASTM standard three-point bend test did not provide a measure of the interlaminar shear strength of a laminated composite.

The primary objective of this research was to ascertain if the observed damage initiation and failure mechanisms of ACC-4 carbon-carbon composites loaded in three-point bending were governed by the contact load as predicted by Copp and Keer, as well as, to define when and where the damage occurred. Since the ASTM-2344 three-point bend test has been employed by many researchers, it was selected as an excellent test vehicle to demonstrate the damage initiation and growth mechanisms activated by this test. Specifically, this paper contributes to the fundamental understanding of the complex damage and failure modes in two dimensional uncoated woven carbon-carbon materials.

EXPERIMENTAL PROCEDURES

A series of ASTM standard three-point bend tests were conducted on warp aligned ACC-4 carbon-carbon composites for length (L) to depth (d) ratios of 4, 5 and 15. These ratios were chosen based upon Copp and Keer's [4] analyses and Whitney and Browning's [3] experiments on graphite/epoxy laminates which showed that a change in the failure mode would occur between the short beam shear (L/d = 4,5) and the long beam geometries (L/d = 15). Figure 1 shows a schematic of the test fixture used. The "failure load" was defined as the ultimate load obtained on the load-displacement curve.

Specimens were examined through a high-magnification microscope and photomicrographs were taken to record the damage. The tested ACC-4 material was X-rayed prior to machining and no internal delaminations or defects were detected. A total of fifteen specimens were prepared, six with L/d = 4, four with L/d = 5, and five with L/d = 15. The specimen cross-sections were rectangular (width-to-depth ratio, w/d = 2) instead of square as recommended by ASTM standards [1]. This was done to allow comparison of the present studies results with Szaruga's [5] four-point bend tests of similar rectangular specimens. The specimens' sides were polished and photomicrographed in the indenter and support regions, and along the neutral axis. Figure 2 shows a typical photomicrograph of the untested specimens. This clearly shows the voids formed during the processing of this material. The void content for these specimens ranged from ten to fifteen percent of the polished surface area.

The specimens were mounted in an Instron machine, then loaded by the motion of the crosshead at the lowest speed (0.05 mm/min) to minimize any catastrophic damage and to allow the maximum time possible for observation of damage and its progression. A Robinson-Halpern Linear Variable Differential Transducer (LVDT) was used to measure the mid-span deflections. The LVDT was mounted inside the support stand and aligned so that the transducer core rod protruded through the support plate and touched the center of the lower surface as shown in Fig. 1.

During the test, the mid-point displacement and applied load were continuously monitored on an x-y plotter to provide a permanent record of the damage history. The maximum loads (P_{max}) were determined by conducting four tests up to failure for each of the L/d ratios. Once again, for these P_{max} tests, specimens were loaded until the applied load reached a maximum value and started to decline. At this point the crosshead motion was reversed and the specimen was unloaded.

The next series of tests involved the loading and unloading of specimens incrementally in order to establish the damage initiation site. Once this threshold was defined, the tests were continued to determine when and where other damage mechanisms became active. For these tests, the specimens were loaded to predetermined levels of the expected ultimate load, and then unloaded. The parameter used to define these predetermined load ratios was $R = P/P_{max}$, where P_{max} was estimated from the previous tests. The specimens were removed from the machine and examined for damage initiation or damage progression which occurred during that increment of testing. Upon completion of each incremental test, all data were corrected for the specimen's actual maximum load.

Measuring the maximum loads ensured that the results were not being affected by the removal and re-insertion of the specimens in the test fixture. The beam mid-point displacements for these incremental tests returned approximately to zero upon unloading whenever the specimens were not significantly damaged. Only when the damaged area became sufficiently large was it possible to observe a permanent deformation. However, this permanent deformation was not quantifiable and it could not be used as a measure of the resulting damage. As a further check of the test results, a continuous incremental test was conducted on a short beam specimen. The beam was loaded to the estimated R value then unloaded to $R \cong 5\%$ and reloaded to the next R value without stopping until the maximum failure load was determined.

DISCUSSION OF RESULTS

A total of 12 specimens, four of each L/d ratio, were loaded to failure and microscopically examined. Figure 3 shows the results of these P_{max} tests and compares them with four-point bend data from similar ACC-4 material obtained recently by Szaruga [5]. Note that the difference in P_{max} is greatest for the L/d = 4 specimens and decreases as the L/d ratio increases. Examination of photomicrographs revealed that extensive damage existed throughout the P_{max} tested specimens. All specimens exhibited an outer ply fracture at either the indenter or support pin contact area. This type of damage is shown in Fig. 4 for a long beam (L/d = 15). Pre- and post-test micrographs clearly showed this cracking occurred during the test and was not caused by cutting and polishing the specimens. However, it was not clear whether the cracks initiated internally or at the surfaces of the beams. In addition, the outer ply fractures did not always penetrate completely through the thickness. For one specimen, the top ply fracture was visible on the back face, but not on the front face. This indicated the damage was three dimensional in nature.

The mode of failure along the neutral axis of the beam was a series of angled shear cracks in both the short and long beam specimens. The microcracks were found in the fill layers, generally contained between the two warp plies in the center portion of the laminate. Examples of the types of cracks that could be found on either side of the indenter load point, in a somewhat symmetric pattern, are shown in Fig. 5. One side of the beam was weaker than the other and eventually the cracks would spread only in the right or left side of the beam. Figure 6 provides an example of these extended shear cracks in the fill layers. The cracks at the neutral axis did not always propagate equally through the thickness of the beams again substantiating the three dimensional behavior of these materials.

The short beam specimens exhibited multiple cracks in adjacent layers of fill fibers proceeding away from the indenter load point and the neutral axis. Microscopic examination revealed that these cracks only appeared in the fill layers and only in the section of the beam between the supports. This was expected since the shear force in the beam should be zero outside the supports. The cracks in the long specimens were generally longer, wider, and more interconnected than those in the short beam specimens.

During the testing of the short beam specimens, no catastrophic failures were ever observed, just a gradual progression of damage as described above. However, during the testing of a long beam specimen (L/d = 15), a sudden catastrophic failure was observed just after the applied load reached the maximum value. This failure was evident by the loud snap that was heard from the specimen and the visible "crack" which appeared along the mid-plane surface of the beam. This loud "snap" and visible surface fracture may lead one to incorrectly assume the test induced a single, long shear crack in the laminate as expected for a three-point bend test. However, microscopic examination of this visible "crack" revealed it was a collection of many smaller angled cracks in the fill fiber layers as described above, and evidenced in Fig. 7. A series of incremental loading tests were conducted to determine the damage initiation and growth mechanisms in these specimens. After each load and unload cycle, the specimens were examined under the microscope to see if any damage was visible. Only one specimen of each geometry (L/d = 4, L/d = 5, and L/d = 15) was tested. The average P_{max} values, for each geometry, were used as the expected failure loads, to calculate selected load ratios of R from 50% to 100%, in increments of 10%. At the completion of each test the R value was corrected for the actual maximum load of that specimen. References to R in this paper refer to this corrected value. The maximum loads attained from the incremental tests were in agreement with the maximum loads obtained in the earlier P_{max} tests. This important observation indicated that removing the specimen from the fixture did not affect the overall strength of the specimen.

The initial damage mechanism in all cases was a fracturing of the outer ply fibers at either the indenter load point or the support points as shown in Fig. 4 for an L/d = 15 and at a load ratio of R = 63%. This initial fracture occurred at a load ratio of R = 46% for the short beam specimens. High magnification views of the neutral axes did not reveal the presence of any shear cracks in either of these specimens.

Upon further loading the next damage modes that occurred were the angled shear cracks in the fill layers. They appeared simultaneously in the neutral axis and indenter regions, just off the load centerline, as was expected from Copp and Keer's analyses [4]. Figure 5 shows these initial shear cracks for an L/d = 15 and at a load ratio of R = 74%. Figure 6 shows the growth of these cracks as they extend through the fill layers for an L/d = 15 and at a load ratio of R = 85%. Figure 7 shows the final failure surface at the maximum load. It is clear that a series of shear microcracks has been generated, however these cracks did not coalesce into a single crack as one might have assumed for this catastrophic failure. In addition, Fig. 8 shows that crack progression was not affected by the presence of large voids in the specimen. Once again, comparison of the incremental maximum failure results with the P_{max} tests results shows no apparent change in the damage mechanisms as a result of the repeated removal, reinsertion, and reloading of the specimen. A schematic representation of the damage progression as described by the above mentioned micrographs is shown in Fig. 9 for a long beam specimen. It should also be noted that the final damage of the continuous incremental load test, in which the specimen was not removed for microscopic examination, was also in agreement with the P_{max} tests and the incremental tests for the short beam specimens.

CONCLUSIONS

These experiments showed that damage and failure mechanisms, in three-point bend tests, of ACC-4 carbon/carbon composites are complex and three dimensional in nature. In all experiments the outer ply fractured at the contact points prior to any internal damage. This fracture occurred at as low as sixty-three percent of the maximum load determined from the P_{max} failure tests for the long specimens and at forty-six percent for the short specimens. Following the outer ply fracture, angled shear cracks formed simultaneously in the fill layer plies around the indenter contact region and the neutral axis of the beam. In short span beams, cracking progressed from the contact point regions and the neutral axis towards each other into the adjacent plies. In the long span beams, cracking occurred in the vicinity of the contact region but did not progress toward the neural axis. The cracks in the neutral axis continued to grow but did not coalesce into a large single crack. Instead, the actual crack that has been associated with the long specimens was merely a series of these smaller microcracks. This work clearly shows that initiation always occurs in the the contact regions even for the long beam geometries. Thus damage initiation and failure mechanisms in the three-point bend test are governed by the indenter contact loads as predicted by Copp and Keer. Therefore, this test does not provide an accurate measure of the interlaminar shear strength of a composite laminate. However, the three-point bend test can provide a useful damage initiation test for the investigator so long as one understands the stress fields present.

ACKNOWLEDGEMENTS

The authors are indebted to Mr. T.G. Fecke of the Components Branch of the Turbine Engine Division of the Wright Research Development Center at Wright-Patterson Air Force Base, Ohio for sponsoring this research.

REFERENCES

1. American Society for Testing and Materials, <u>1985</u> <u>Annual</u> <u>Book</u> <u>of</u> <u>ASTM</u> <u>Standards</u>, Section 15, Volume 15.03, Philadelphia, PA., pp. 55-58.

2. Whitney, J. M. and Browning, C. E., "On Short Beam Shear Tests for Composite Materials," <u>Experimental</u> <u>Mechanics</u>, Vol. 25, 1985, pp. 294 -300.

3. Whitney, J. M., "Elasticity Analysis of Orthotropic Beams Under Concentrated Loads," <u>Composites</u> <u>Science</u> <u>and</u> <u>Technology</u>, Vol.22, 1958, pp. 167-184.

4. Copp, P. D. and Keer,L. M., "Stress Fields in Two Orthotropic Layers Loaded by a Rigid Indenter," Submitted to the <u>Journal</u> <u>of</u> <u>Applied</u> <u>Mechanics</u>.

5. Szaruga, S. L., "Four Point Flexure (Shear) Testing of Structural Carbon-Carbon Composites." An open address to local Scientists and Engineers, Air Force Wright Aeronautical Laboratories, Wright-Patterson AFB, OH, 12 September 1988.

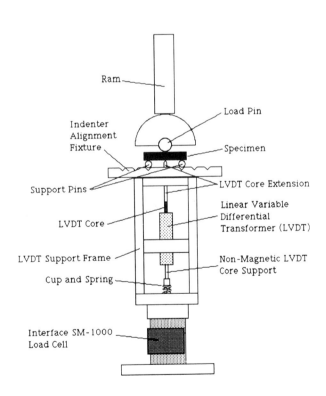

Fig. 1 Schematic drawing of the experimental test fixture.

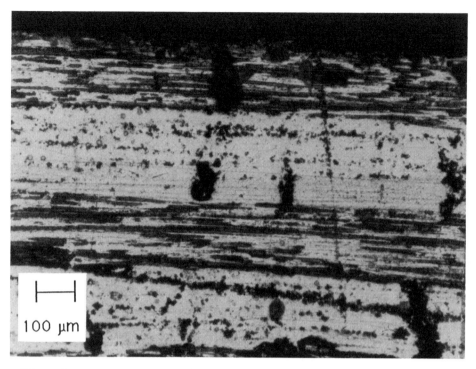

Fig. 2 Undamaged ACC-4 specimen, indenter contact area,
L/d = 15, (100X).

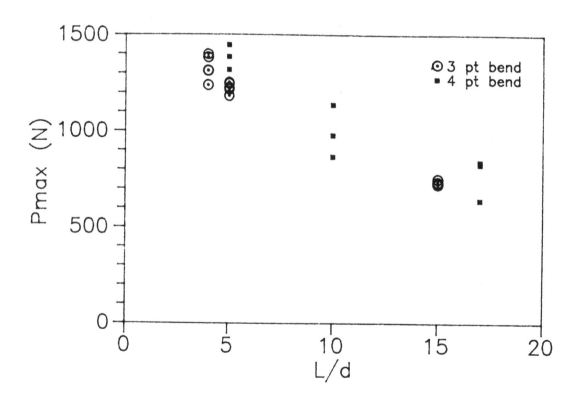

Fig. 3 Test results, maximum load vs L/d ratio for
three-point and four-point bend tests.

Fig. 4   Initial ACC-4 damage mode, outer ply fracture at
the support pin, L/d = 15, R = 63%, (400X).

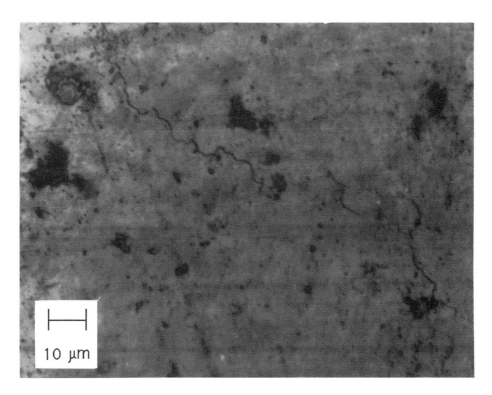

Fig. 5   Second ACC-4 damage mode, angled shear cracks along
the neutral axis, L/d = 15, R = 74%, (1000X).

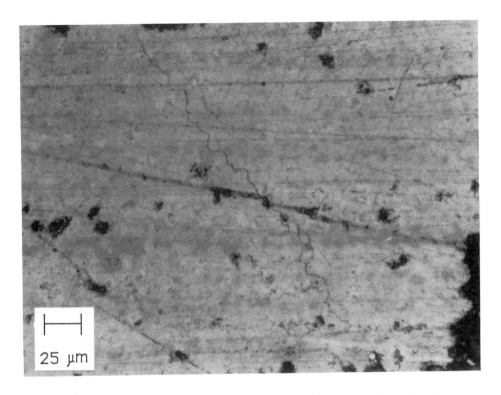

Fig. 6   Growth of shear cracks in ACC-4 composite laminates
along the neutral axis, L/d = 15, R = 85%, (400X).

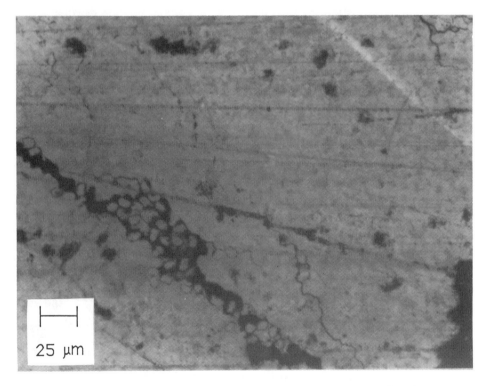

Fig. 7   Catastrophic failure in ACC-4 composite laminates
along the neutral axis, L/d = 15, R = 100%, (400X).

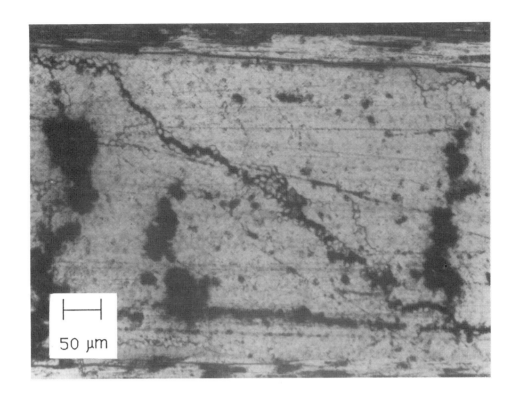

Fig. 8 Catastrophic failure in ACC-4 composite laminates
along the neutral axis, L/d = 15, R = 100%, (200X)

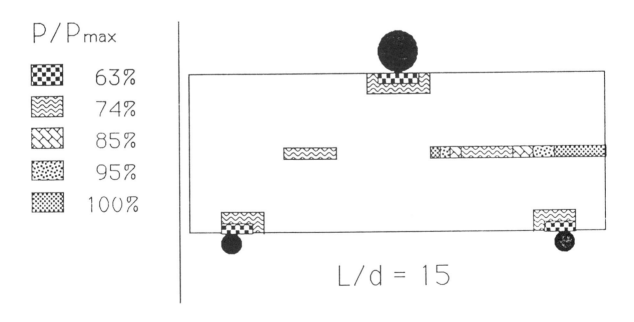

Fig. 9 Schematic history of the experimental damage
progression in ACC-4 composite laminates.

Micromechanics of 3-D Braided Hybrid MMC

CHARLES S. C. LEI AND FRANK K. KO
Fibrous Materials Research Center
Drexel University
Philadelphia, Pennsylvania 19104

ALBERT S. D. WANG
Dept. of Mechanical Eng.
Drexel University
Philadelphia, Pennsylvania 19104

ABSTRACT

The macroscopic elastic behavior of 3-D braided hybrid composites is characterized on the basis of a micromechanical analysis of a unit cell structure. Treating a 3-D braided hybrid composite as an assembly of individual unit cells idealized as a pin-jointed truss in the shape of a brick, the Finite Cell model is established based on the principle of virtual work and structural truss analysis. The model is evaluated using the tensile properties of SiC/AL composites.

INTRODUCTION

Silicon carbide aluminum composites have received a great deal of attention due to their high environmental resistance, mechanical properties such as the ability to mix well with molten aluminum, reliability at high temperatures, and low raw material cost. 3-D braided SiC/AL composites have been demonstrated to be a class of delamination free composites distiguished by high levels of resistance to crack propagation and superior damage containment capabilities. Longitudinal lay-in reinforcement provides a strengthening effect in 3-D braided hybrid SiC/AL composites.

The most suitable SiC fiber for MMC is Avco SCS-6 which is a SiC fiber formed by CVD on a carbon filament. Although this fiber displays good thermal strength retention, the diameter of the fiber (142μ) makes it unsuitable for 3-D textile processing. Accordingly, a small diameter (14μ) ß-SiC yarn (Nicalon®) was chosen as the fibrous material to form the 3-D architecture. By combining the relatively flexible, albeit inferior, Nicalon fiber with the rigid SCS-6, toughness and through thickness properties can be improved with the 3-D structure and axial stiffness can be retained with 0° "triaxial" braiding yarns (SCS-6).

In this study, 3-D braiding was used as the reinforcing architecture with triaxial (0°) lay-ins incorporated into the preform. Based on the fabrication technique, geometric descriptions of the fiber/yarn structural geometry can be formulated. Figure 3 illustrates the general form of the 3-D braided preform. In particular to this program, the geometric and subsequent mechanical effects of placing triaxial yarns is of relevance. Based on the fabrication technique and hybridization levels, schematic cross-sections of the braided composites can be generated. Figures 1 through 3 illustrate the ideal appearance of the hybrid SiC cross-sections for 100/0, 75/25, and 50/50 ratios of Nicalon/SCS-6 respectively. In order to fully explore the potential of these new material systems, an analytical framework is needed to link fiber architecture and material properties to composite properties. Several analytic models have been developed to characterize the elastic moduli and structural behavior of 3-D braided composites.

Ma, Yang and Chou [1] assumed the yarns in a unit cell of a 3-D braided composite as composite rods, which form a parallel pipe. Strain energies due to yarn axial tension, bending and lateral compression are considered and formulated within the unit cell. By Castigliano's theorem, closed form expressions for axial elastic moduli and Poisson's ratios have been derived as functions of fiber volume fractions and fiber orientations.

Ma, et.al. also developed a "Fiber Inclination Model" according to the idealized zig-zagging yarn arrangement in the braided preform[2]. They assumed an inclined lamina as a representation of one set of diagonal yarns in a unit cell. In this way, four inclined unidirectional laminae form a unit cell. Then, by the employment of classical laminate theory, the elastic moduli can be expressed in terms of the laminae properties.

From a 'preform processing science' point of view, Ko et al.[3] developed a "Fabric Geometric Model" using similar assumptions. The stiffness of a 3-D braided composite was considered to be the sum of stiffnesses of all its laminae. A maximum strain energy criterion was used to determine the failure point for each lamina by taking bending stress on yarn crossover into consideration. The stiffness matrix forms a link between applied strains and the corresponding stress responses. Throughout this analysis, the stress-strain characteristics of the composite are determined.

The results of elastic properties from the above three models can be used as input to a generalized finite element program in order to analyze more complex shaped structures. By doing so, the 3-D braided composite has to be treated as effective continuum, and the unique characteristics of each individual yarn and matrix are smeared out. In complex structural shapes such as I-beams, turbine blades, the final structure often consists of several types of fiber architecture. With these complex fiber architecture systems, the effective continuum concept can no longer provide accurate description. Therefore, a finite cell model (FCM) is presented, which can accommodate structures with variable unit cells and provide a link between microstructural design and macro-structural analysis.

FINITE CELL MODELLING

The FCM is based on the concept of fabric unit cell structure and structural truss analysis. The fiber architecture within the composite is considered as an assemblage of a finite number of individual structural cells with brick shape. Each individual cell is the smallest representative volume from the fibrous assembly. The unit cell is then treated as a space structure with the endowed representative architecture, rather than a material with a set of effective continuum properties.

The key step in the formulation of the problem is the identification of the unit cell's nodal supports, similar to the nodal points of a conventional finite element. In this model, the yarns are assumed to travel along the diagonals in a unit cell and are treated as pin-jointed two-force truss members. By treating a unit cell specifically as a 3-D space truss , a 3-D truss finite element technique may be employed for the mechanistic analysis.

In order to include the effect of matrix, which is subjected to tension or compression under the deformation of yarns, the matrix is assumed to act as truss members, connecting the two ends of a given set of yarns in the unit cell as shown in Figure 4. In the unit cell, the nodes, or the ends of yarns, are pin-jointed with three degrees of freedom in translation. Hence, the matrix plays a role in restricting the free rotation and deformation of yarns. There are a total of 24 degrees of freedom in a four diagonal yarn unit cell. In this analysis, the interaction at the yarn interlacing and bending effect of yarns are not considered.

Let a_{ij} represent the value of member deformation q_i caused by a unit nodal displacement r_j. The total value of each member deformation caused by all the nodal displacements may be written in the following matrix form:

$$\{q\} = [a]\{r\} \tag{1}$$

where [a] is called the displacement transformation matrix which relates the member deformations to the nodal displacements. In other words, it represents the compatibility of displacements of a system.

The next step is to establish the force-displacement relationship within the unit cell. For a pin-connected truss, the member force-deformation relationship can be written as:

$$[Q] = [K']\{q\} \tag{2}$$

where:

$$AE/L$$
$$AE/L$$
$$[K'] = \qquad\qquad AE/L$$

The principle of virtual work states that the work done on a system by the external forces equals the increase in strain energy stored in the system. Here, the nodal forces can be considered as the external forces of the unit cell. Therefore, if $\{R\}$ represents the nodal force vector, it follows that

$$\{\underline{r}\}^T\{R\} = \{\underline{q}\}^T\{Q\} \qquad\qquad (3)$$

where $\{\underline{r}\}$ and $\{\underline{q}\}$ are virtual displacement and deformation, respectively. From EQs.(1) and (2), the following equations can be derived:

$$\{R\} = [K]\{r\} \qquad\qquad (4)$$

where: $\{R\}$ = nodal forces
$\qquad [K] = [a]^T[K'][a]$ = stiffness matrix of the unit cell
$\qquad \{r\}$ = nodal displacements

Using Equation (4), the nodal force and the nodal displacements of a unit cell are related. Thus, for a structural shape which consists of a finite number of unit cells with a specific assemblage pattern, a system of equations for the total structural shape can be assembled using the individual cell-relations following the finite element methodology.

For the analysis of 3-D hybrid composites with 0° longitudinal lay-in yarns, two possible models are suggested to take account of the effect of the longitudinal reinforcement, as shown in figure 5. The first model assumes that the longitudinal yarns penetrating through the center of the unit cell with matrix-bars connecting the ends of the longitudinal yarns. The second model is to distribute the elastic properties of the 0° lay-in yarns into four longitudinal matrix-bars. The resultant properties of the four longitudinal composite bars are obtained by rule of mixture between lay-in yarns and matrix-bars. For the first model, he fomulation of the stiffness matrix is similar to that of 100% braided composites except that the number of the nodes in each unit cell is two larger. As for the second model, only the properties of the four longitudial members are different, but formulation is as just described as above. With basic parameters in a unit cell, such as yarn elastic modulus, fiber volume fraction, yarn orientation and unit cell dimension fully characterized, the accruacy of the two models to predict the structural response of composites will be evaluated with experimental data.

EXPERIMENTAL EVALUATION

Simple rectangular coupons of the 3-D braided hybrid Nicalon/Avco aluminum matrix composite were fabricated and characterized by tensile testing. Figure 6 shows the experimental stress-strain relationships for these materials. In this figure, the ultimate strains of both cast aluminum-6061 and 100% braided composite are low as expected. However, the more longitudinal Avco filaments are embedied in the composite, the more dominant effect of the Avco filaments the composite shows. The nonlinear behavior of the MMC may be attributed to the local filament breakage of lay-in yarns, the nonlinear structural response and the nonlinear behavior of the aluminum matrix. However, the contribution of each cause is not well known yet; in this sense the FCM does not include the analysis of this phenomenon. In this study, only the elastic behavior of the material system , or, the initial modulus, will be predicted.

The designed value of 35% of fiber volume fraction was used for the numerical computation. The dimension of a unit cell is determined from the measurements. Since the dimensions of a unit cell are considered to be the center lines of members of the unit cell, part of the bars lie outside the unit cell. Thus, an averaging method for the determination of the cross-section areas of the bars was used. For a 100% braided composite with a unit cell dimension of HxWxT, the area of a fiber-bar can be obtained as $A_f = 0.35HWT / 4(H^2+W^2+T^2)^{1/2}$; the area of a matrix-bar can be expressed as $Am = 0.65HWT / 4(H+W+T)$. As for hybrid composite with Nicalon/Avco ratio being m/n, the effective area of Nicalon yarns is $m(0.35HWT / 4(H^2+W^2+T^2)^{1/2})$, while the Effective area of Avco bundle is $n(0.35WT)$. In model (a), the effective area of matrix-bars is $0.65HWT / 4(H+W+T +(W^2+T^2)^{1/2})$; while in model (b), the effective area of the four longitudinal composite bars is determined by rule-of-mixture over the longitudinal aluminum-bars and Avco-bars. The formula of the rule-of-mixture is as follows:

$$A_{Aluminum\text{-}bar}E_{Aluminum\text{-}bar} + A_{Avco\text{-}bar}E_{Avco\text{-}bar} = AE_c$$

where: $A = A_{Aluminum\text{-}bar} + A_{Avco\text{-}bar}$

Accordingly, the elastic properties used for the unit cell are as follow:

$E_{Avco} = 40$ msi ; $E_{Nicalon} = 28.5$ msi ; $V_f = 0.35$
$E_m = 7.95$ msi ; $V_m = 0.65$

Table 1 shows the unit cell dimension and effective area of each type of bar.

Since each cell has an similar geometry and boundary conditions, it is sufficient to model only one element, as shown in Figure 7. Equal load was applied to the nodes at one end of the cell to simulate the condition of a coupon under uniform tension.. The applied load was divided into several load steps on account of the possible nonlinear load-deformation behavior due to geometrical conformation.

The predicted stiffness of the composites tended towards a higher value than experimental results. This can be attributed to the use of fiber data as an input for our prediction. In order to reflect the fiber breakage and degradation during manufacturing, the use of yarn data may be more appropriate. The comparison of experiment with theory shows a reasonable agreement. Table 2 summarizes the results and compares experimental results with predicted results. As seen in this Table, model (a) predicts a lower modulus with higher percentage of Avco filaments. This contradicts the experimental results. The reason is that to put the longitudinal yarnss along the central line of a unit cell makes the whole structure respond non-homogeneously. Therefore, for more Avco filaments aligned along the central line, the effect shows up. As for the FGM, the assumption of unidirectional laminate in the yarn orientation results the overall anisotropic properties of 3-D braided system. Hence, the coupling effect between extension and bending is accounted in the computation procedures. However, extended studies have to be done on this subject.

CONCLUDING REMARKS

Two finite cell models have been presented to predict the mechanical behavior of 3-D braided hybrid composites. The first model, which assume lay-in yarns travelling along the central line of a unit cell, has been shown unsuitable to describe the behavior of the unit cell under uniform tension. By distributing the reinforcement effect of lay-in yarns into the longitudinal matrix-bars, the second model shows a reasonable agreement with experimental results. Therefore, by appropriate choice of yarn mechanical properties and precise determination of dimension of a unit cell, the 3-D truss FCM has been shown to be an adequate model for a 3-D braided composite for a first approximation.

In a 3-D braided composite, the yarns actually experience bending moments throughout the unit cell under tensile loading. The present model will be extended to include the bending effect by replacing the pin-jointed truss as a stiffer frame structure. It should be pointed out that the fiber geometry of a unit cell along the boundary of a specimen is slightly different from the one at the center. In order to more precisely characterize the load-deformation relationship of the whole composite, especially at the corner of a complex shape 3-D braid composite, a few different unit cell may be introduced within an analysis.

ACKNOWLEDGEMENTS

This study is supported in part by the AirForce Office of Scientific Research.

REFERENCES

1. Ma, C. L., Yang, J. M. and Chou, T. W., "Elastic Stiffness of Three-Dimensional Braided Textile Structural Composites," Composite Materials: Testing and Design (Seventh Conference), ASTM STP 893, J. M. Whitney, Ed., American Society for Testing and Materials, Philadelphia, 1986, pp. 404-421.

2. Yang, J. M., Ma, C. L. and Chou, T. W., "Fiber Inclination Model of Three-Dimensional Textile Structural Composites," J. of Composite Materials, Vol 20, 1986, pp 472-484.

3. Ko, F. K., Pastore, C. M., Lei, C. and Whyte, D. W., "A Fabric Geometry Model for 3-D Braid Reinforced FP/AL-Li Composites," Competitive Advancements in Metals/Metals Processing, SAMPE Meeting, Aug. 18-20, 1987, Cherry Hill, NJ.

4. Ko, F. K. and Pastore, C. M., "Structure and Properties of an Integrated 3-D Fabric for Structural Composites," Recent Advances in Composites in the United States and Japan, ASTM STP 864, J. R. Vinson and M. Taya, Eds., American Society for Testing and Materials, Philadelphia, 1985, pp.428-439.

TABLE 1. Properties of unit cell (area in in^2; modulus in Mpa)

Ratio of Nicalon/Avco	A_{matrix}	$A_{Nicalon}$	model (a) E_{Avco}	A_{Avco}	model (b) E_{C-rod}	A_{c-rod}
100/0	.0008977	.0004613				
75/25	.0009656	.0004561	276.	.0006366		
75/25	.0009656	.0004561			87.2	.00113
50/50	.001498	.0004524	276	..0019445		
50/50	.001498	.0004524			104.	.001984

TABLE 2. Comparison of Experiment and Theory of Hybrid SiC /Al MMC
(modulus, Mpa; $V_f = 35\%$)

Ratio of Nicalon/Avco	EXPERIMENTAL	FGM	FCM model(a)	model(b)
100/0	88.3	98.87	91.04	
75/25	95.6	108.3	123.6	97.45
50/50	102	117.4	109	113.45

Figure 1. Schematic Cross-Section of 100% Braided Preform

Figure 2. Schematic Cross-Section of 75/25 Braid/Longitudinal
3-D Braided Preform

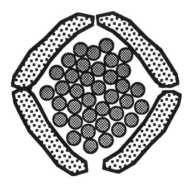

Figure 3. Schematic Cross-Section of 50/50 Braid/Longitudinal
3-D Braided Preform

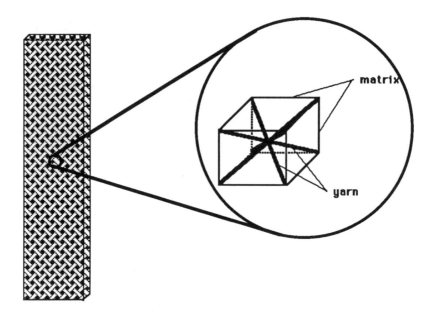

Figure 4. Schematic of a 3-D Braided Composite Unit Cell Geometry

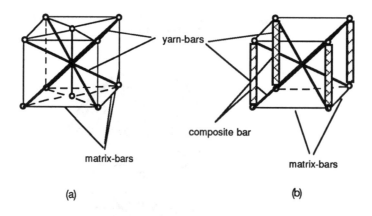

Figure 5 Two approaches for analyzing a unit cell of hybrid structure.

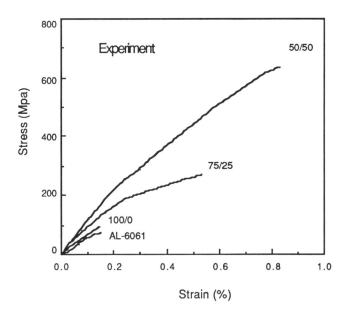

Figure 6. Experimental Results of hybrid MMC.

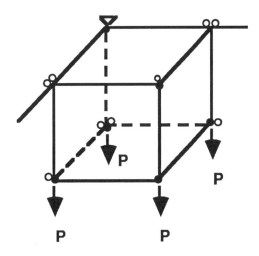

Figure 7. Boundary and Loading Conditions of a Unit Cell.

Performance Evaluations of Oxidation-Resistant Carbon-Carbon Composites in Simulated Hypersonic Vehicle Environments

D. M. BARRETT, W. L. VAUGHN, H. G. MAAHS, C. W. OHLHORST AND R. H. MARTIN

ABSTRACT

Because of their attractive high temperature structural properties, carbon-carbon composites are being considered as hot structural materials for advanced hypersonic vehicle airframes. Before such an application can be realized, however, adequate oxidation protection must be demonstrated for these carbon-carbon composites. This paper focuses on this issue of oxidation protection. A number of coated carbon-carbon specimens developed for aeropropulsive environments were tested in simulated aerostructural environments. Maximum test temperatures were 2800°F, and pressures ranged from 0.03 atm to 1 atm. The specimens were exposed intermittently to high humidity between tests. The test medium was air.

Test results concerning the effects of coating thickness, substrate architecture, coating composition, and substrate surface preparation on the oxidation resistance of coated carbon-carbon composites in a representative hypersonic vehicle environment are discussed. Based on the data obtained, it is concluded that both surface preparation techniques and coating chemistry have profound effects on the coating adherence and longevity of these composites in hypersonic vehicle environments.

INTRODUCTION

Low density, high specific strength and stiffness, relatively low coeffcients of thermal expansion, retention of mechanical properties at high temperature, and chemical and thermal stability are all desirable characteristics of high temperature structural materials for hypersonic vehicle airframe

D. M. Barrett, H. G. Maahs, C. W. Ohlhorst, National Aeronautics and Space Administration, Applied Materials Branch, Hampton, Virginia 23665-5225
W. L. Vaughn, Planning Research Corporation, Hampton, Virginia 23665-5225
R. H. Martin, National Research Council Research Associate, Hampton, Virginia 23665-5225

applications. With the notable exception of chemical stability, specifically oxidation resistance, carbon-carbon composites have all of these characteristics. Accordingly, oxidation protection systems (OPS) for carbon-carbon structural components are undergoing development which will allow coated carbon-carbon materials to be used as hot structural materials.

A number of OPS concepts have been developed for carbon-carbon composites. Early developmental coatings were based on SiC as a coating material.[1] Evolution of the SiC coatings resulted in boron additions to the SiC.[2] The boron provides an oxide sealant which forms in-situ during oxidation and promotes flow at intermediate temperatures to seal coating cracks formed during processing. Since depletion of the limited amount of boron leads to accelerated oxidative failure, boron and other inhibitors were incorporated within the substrate matrix to enhance the longevity of these materials in an oxidizing environment.[2,3] However, even with substrate inhibitors and boron additions to the SiC coatings, insufficent longevity was obtained from SiC-based materials exposed to cyclic oxidizing environments.[4] Accordingly, B_4C-based materials were added as coating underlayers to increase the reservoirs of available boron.[5] These larger boron reservoirs could extend the lifetime of available oxidation protection systems.

Oxidation of boron and boron carbide occurs too rapidly to use these materials by themselves as a coating in a cyclic, oxidizing environment. Therefore, SiC overcoats must still be utilized as the primary oxidation defense. Because the SiC overcoats are cracked, the boron bearing inner layers will oxidize. In high-temperature aerostructural environments, the oxidation products of the boron compounds tend to be volatilized at the higher temperature, lower pressure regimes of the flight profile. This leads to separation of the SiC from the underlying boron compounds followed by rapid erosion of the boron layer, followed in turn by substrate oxidation.

Increased coating thickness increases the lifetime of coated carbon-carbon.[6] Thicker coatings tend to provide a larger boron reservoir, more completely cover the substrate, are more forgiving to variations in the coating process and allow more processing defects. As will be demonstrated by the results presented in this paper, these adjustments provide an increase in specimen lifetime.

Figure 1 is a plot of the linear thermal expansions of SiC, Si_3N_4, B_4C and LTV Aerospace and Defense Company's (LTV's) structural, 2-D carbon-carbon composite designated ACC.[7,8,9,10] A mismatch exists in the coefficents of thermal expansion (CTE) between SiC ($3x10^{-6}$ $°F^{-1}$) and ACC ($1x10^{-6}$ $°F^{-1}$; in-plane). This mismatch causes crack formation in the coating when the coated composite is cooled from the elevated processing temperature to room temperature. On small specimens (1 by 3 by 3/16 inches) typical maximum crack widths of 10 micrometers have been observed in addition to a number of smaller sized cracks.[11] The calculated percentage crack area of a SiC coating on the face of an ACC substrate is also shown in Figure 1, assuming a coating deposition temperature of $3000°F$.[12] Science Applications International Corporation (SAIC) and other organizations are modelling crack patterns of various coated carbon-carbon composite systems.[13]

The CTE of SiC is much higher than the in-plane CTE of 2-D carbon-carbon composites, as shown in Figure 1. However, the through-the-thickness CTE more closely matches the CTE of SiC. Hence, at least theoretically, coatings on the edges of 2-D carbon-carbon composites should perform better than coatings on the faces, although this has not been seen in our testing.[14] Indeed, these variations appear to have little effect on the oxidation performance of coated carbon-carbon composites.

Externally applied glazes and sealants were added[15,16] to further slow the deterioration of the boron constituents of the coating system by providing a relatively involatile crack-filler that will survive the high temperature, low pressure regimes.

As will be shown in this paper, further increases in material lifetime can result from special surface preparation techniques (SPT). SPT details are generally considered to be proprietary information by the OPS producers; however, macroscopic evaluation of the surface morphology and post-oxidized cross sections clearly show a modification of the surface geometry. These modifications can affect the cracking patterns of the coating which in turn can affect the lifetime of the material.

In the present study, specimens with a wide range of thickness, coating composition, substrate architecture, and surface preparation were exposed to simulated hypersonic vehicle mission cycles. Results from these exposures are presented and discussed in this paper.

EXPERIMENTAL

In order to assess the performance of coated carbon-carbon composites in the high-temperature airframe environment, numerous specimens were obtained from a variety of vendors as shown in Table 1. Specimen geometry varied somewhat, but was generally on the order of 0.5 inch by 1.0 inch by 0.1 inch thick. Some of these specimens were fabricated as long as three years ago and hence, do not necessarily represent current state-of-the-art materials. However, they do provide valuable baseline data.

Since advanced aerospace materials may be designed to operate on both cruise and orbital vehicles,[17] representative conditions of temperature and pressure for both of these missions were selected from calculations for many possible missions. These are shown in Figure 2. During cruise missions, maximum temperatures will be much lower than for orbit missions and an extended period of time will be spent at intermediate temperatures. Cruise pressures will not be as low as orbital mission pressures, although they will be significantly lower than one atmosphere. The highest temperatures encountered will be on ascent to and entry from orbit and will be experienced for relatively short durations. Low pressures will be encountered while the vehicle is at its highest temperature. The conditions shown in Figure 2 are not representative of any particular component but were chosen to include some of the extremes which may be encountered. Precise conditions will depend on mission trajectory, vehicle velocity, location on the vehicle, surface properties (emittance, catalytic activity, etc.), and other factors.

All specimens were tested in the NASA-Langley Multiparameter Environmental Simulators, a photograph of which appears in Figure 3. The facility consists of three separate test chambers, all computer-controlled and capable of simultaneously testing up to eighteen specimens. Maximum temperature capability is 3000°F at pressures ranging from 1.3×10^{-8} up to 1.0 atmosphere. Additionally, loads to 2000 pounds may be applied in tension to the specimens. Atmospheric composition may also be varied.

A testing sequence similar to that shown in Figure 4 was used for oxidative testing. Since it is well established that moisture can significantly degrade coating performance,[12,18,19] and because of the limited number of test specimens available for the present study, all tests were performed on humidity-exposed specimens. The primary source of this degradation is hydrolysis of borate-based glasses in moist air.[20] These glasses typically form from boron-based constituents of the oxidation protection system.

Initially the specimens were dried (in a vacuum at 250°F for 16 hours) to obtain a baseline dry weight. The specimens were then exposed to an 80°F, 90-percent-relative-humidity environment for a minimum of three days before the first oxidation test. A two-day testing sequence was used as depicted in Figure 4. The first days testing consisted of two cruise simulation cycles. Specimens were weighed after removal from the humidity chamber and after each cycle. Following the second cruise cycle, the specimens were re-exposed to the humid environment overnight. The second day's testing consisted of a single ascent simulation cycle followed by two entry simulation cycles. The specimens were weighed after each cycle. After completion of the second entry cycle, the specimens were again exposed to humidity overnight. Thereafter, a repeat of the two-day sequence was conducted until the specimens lost 75 g/m^2 of their dry weight (based on the total surface area of the specimen). Figure 5 is a plot of typical results obtained from a specimen tested in this fashion.

COATING CHEMISTRY AND THICKNESS

Specimens 1 through 6 (Table 1) were tested in order to provide baseline data on their particular coating system. No externally applied sealants or glazes were added to the coatings. Referring to Table 1, the Type III material is a SiC conversion coating which is not theoretically dense, while the denser Type IV is a boron-doped SiC conversion coating, and the Type VI is a Type IV overcoated Type III which has undergone an additional resin densification step to provide a carbon source for the formation of the Type IV SiC. The Type IV plus CRT, Type IV plus MID, and Type VI plus CRT are all chemically vapor deposited overcoatings deposited onto the previously described Type IV and Type VI conversion coatings. All coatings, with the exception of the Type IV plus MID, were deposited with the in-plane direction of 2-D carbon-carbon composite constituting the largest area of the substrate surface. The Type IV plus MID was deposited with the through-the-thickness direction of the substrate constituting the largest area of the substrate surface. This substrate had been cut from a larger block of thick substrate.

Variations in coating thickness on the different surfaces of the specimens can occur as shown in Figure 6. Hence, a number of different ways exist to measure coating thickness. Three obvious ways, all based on measurements of a number of points from a cross-sectioned specimen are: 1. Measure the coating thickness and take an average based on each side of the specimen; then weight these averages as a function of lineal distance to obtain an average coating thickness over the entire specimen. 2. Measure and average the coating thickness on each side of the specimen, and select the thinnest side to be representative of the "weakest" surface of the specimen. 3. Determine the thinnest spot on the specimen and use this as the "weakest" region of the specimen. Figures 7a, 7b, and 7c display the results obtained when using these three different approaches to defining coating thickness.

The first approach is felt to be lacking because it is not indicative of a large area 'weak link' in the protection system. The third approach is considered impractical because it would require extensive, destructive cross-sectioning of the specimen to determine the thinnest spot in the entire coated specimen. Therefore, the second approach (Figure 7b) was adopted.

From Figure 7b it is obvious that the performance of the several coatings does not fall on a straight line. This is probably due primarily to variations in the chemistry of the coating system, but other factors may play a role. As can be seen, the results lie in two distinct regimes. The first regime, roughly bounded by two dashed lines, consists of pure SiC coatings. The second regime, which appears as a crosshatched area, includes SiC coatings that contain between one and ten weight percent boron.[21] Figure 8 depicts a microprobe analysis of a sectioned Type IV (boron-containing) specimen. A number of spot analyses were performed for the elements boron, carbon and silicon. From these spot analyses, a quantitative measure of the concentration of these elements was obtained and plotted as a function of distance through the coating.

The boron-containing coatings clearly perform better in the present test environment than do the pure SiC coatings (see Figure 7b). Additionally, it would appear reasonable that the higher the boron concentration in the coating system, the better that coating would perform, at least up to some concentration. However, the total amount of boron in these specimens (or other specimens in Table 1) has not been accurately determined. This uncertainty is at least partially responsible for the variations shown in these plots (Figures 7a, 7b and 7c). Furthermore, the one coating deposited on the through-the-thickness direction of the carbon-carbon composite (Type IV + MID) does not show any appreciable improvement in performance despite the change in the CTE match of the coating and the substrate. Because of the limited data available, it is not possible to discern the effect of the chemically vapor deposited overcoats on specimen performance.

COEFFICENT OF THERMAL EXPANSION (CTE) MISMATCH

A second series of specimens tested had MOD IV coatings (see Table 1). Some of the specimens had been fabricated by San Fernando Labs (SFL) with the standard MOD IV coating chemistry but had thinner coatings than the typical thickness of fifteen mils. The SFL coating technology has since been purchased and improved by SAIC. Using this technology as a basis, SAIC deposited an amorphous alpha Si_3N_4 modified MOD IV coating to permit investigation of the difference in performance between SiC and Si_3N_4 coatings.

The CTE of Si_3N_4 more closely matches that of the carbon-carbon substrate than does the CTE of SiC (Figure 1) and, hence, Si_3N_4 coatings may be expected to show better performance than SiC coatings. Specimens supplied by SAIC were tested in the NASA Langley Multiparameter Environmental Simulators. The results of the exposures on the MOD IV SiC and Si_3N_4 are shown in Figure 9. These results show no significant differences in performance between the amorphous silicon nitride and crystalline silicon carbide coatings tested. Although other factors differ between these two systems, the implication seems to be that the better CTE match of Si_3N_4 to the substrate affords no major improvement over typical SiC coatings in terms of oxidation performance.

EXTERNALLY APPLIED GLAZES

Specimens with and without externally applied glazes were received from Chromalloy Research and Technology Division (CRT). These include 2-D and 3-D angle interlock substrate materials, with and without HITCO-applied conversion coatings, and with and without surface preparations (see Table 1).

Test results for five of these specimens are shown in Figure 10. These specimens had identical coating systems chemically vapor deposited by CRT onto Rohr inhibited-substrate carbon-carbon materials. The chemically vapor deposited coatings were multilayer in construction and consisted of nonstoichiometric boron carbide\silicon carbide ($B_xC\backslash SiC$) with an RT42A closeout coating. RT42A is a CRT designation for a nonstoichiometric, silicon-rich silicon carbide. The composition of this material is 63 to 79 mole percent silicon (80 to 90 weight percent silicon), the balance being carbon. Two of the specimens had no glaze and three had externally-applied glazes. The externally applied glazes appreciably increased the oxidative lifetime of these materials, presumably by retarding the ingress of oxygen to the substrate through the coating cracks.

Figure 11 schematically depicts the presumed mechanistic difference between unglazed and glazed materials. The oxidative protection of the unglazed system is provided by the boron constituents and other oxides available in the system. This boron can be depleted through volatilization of the boria formed during high-temperature oxidation. Boria also hydrates readily to form orthoboric acid. Hence, the available boron reservoir can limit the lifetime of the material. An externally applied glaze can seal the coating cracks prior to the formation of boria from atmospheric oxidation. When diffusion of oxygen occurs through the oxide glaze to the underlying coating, boria counterdiffusion and mixing with the glaze would be expected to slow the depletion of the boron reservoir.

SUBSTRATE ARCHITECTURE

Figure 12 shows the performance behavior of specimens with identical coatings but having different substrate architectures. Two of the specimens had a 3-D angle interlock substrate whereas the others had the typical 2-D substrate. Although differences exist in the oxidative behavior of these specimens, the variations are small, in some cases smaller than the performance between similar specimens. One reason for this similarity in performance is that the fibers in the through-the-thickness (z) direction have very little effect on the in-plane CTE. Therefore, cracking of the coatings due to CTE mismatch will occur to the same extent regardless of substrate architecture. However, it is reasonable to presume the fibers in the z-direction could cause local stress concentrations on the substrate surface, which could induce local secondary cracking in the coating. However, if this does occur, Figure 12 shows the effect to be minimal.

SURFACE PREPARATION

Two specimens exhibited longer oxidation lives than those discussed heretofore; these specimens contained CRT's proprietary surface preparation technique. As can be seen in the photomicrograph in Figure 13, this preparation technique involves, at a minimum, a modification of the surface topography of the substrate. Although a thorough analysis of this relatively new technique has not been performed, limited SEM observation reveals that the coating experienced transverse (through-the-thickness) cracking as well as cracking parallel to the substrate surface.

Residual in-plane tensile stresses in flat brittle coatings deposited on flat-surfaced substrates cause steady state transverse cracking, thus exposing the substrate to the atmosphere. Crack propagation through the coating may be determined by calculating the strain energy release rate, G[22]. The residual stress state in the coating may be altered by altering the surface geometry of the substrate. Hence, the failure mode in the coating may be changed along with the oxidative performance of that specimen.

To better understand the resulting residual stress distribution caused by a surface preparation such as CRT's a simplified finite element analysis was conducted with NASTRAN[23]. The coating was modelled as a simple two dimensional repeating cell, with appropriate boundary conditions applied. The finite element mesh used is shown in Figure 13. NASTRAN's CQUAD4 elements were used with membrane, bending and transverse shear capabilities. Both the coating and the substrate were modelled (for purposes of simplification) as isotropic materials, with a CTE ratio of 3.0 (coating to substrate). Figures 14 and 15 depict the in-plane and normal stresses, respectively, as a function of distance along the coating, s, normalized by the total coating length, D, for different locations through the thickness. In both figures stress, σ , has been normalized by:

$$\sigma_0 = E_s \, \alpha_s \, \Delta T \qquad\qquad (1)$$

where E_s and α_s are the modulus and CTE of the substrate, respectively, and ΔT is the temperature difference from the stress-free state. For this problem, 2300F° was used for this temperature difference. The in-plane stresses cause transverse cracking and the normal stresses cause in-plane splitting parallel to the substrate. Figures 14 and 15 show that the highest stresses are in the trough region whereas the peak region is largely unstressed. At a s/D ratio of 0.0 in the trough there are only in-plane stresses which are largest at the surface. Therefore, transverse cracking may be expected at this location as seen in Figure 13. It can be determined from Figure 15 that the normal stresses reach a maximum on the slope of the trough, at a s/D ratio of about 0.4. These normal stresses are caused by the moment created by bending the substrate surface during the cooldown following processing and promote the in-plane splitting observed in Figure 13. As a result of the surface preparation, transverse cracking has been restricted to the trough regions. The location of the cracks is deeper within the coating system and therefore more protected. Hence, an improvement in oxidative behavior would be expected and, as seen in Figure 16, does occur.

Figure 16 shows the influence of the surface preparation on the performance of these materials. Three specimen types, those with no surface preparation, those with a surface preparation and a conversion coating and those with just a surface preparation are shown in this plot. Substrate material was HITCO's CC137E and all specimens had CRT's multilayer coating of $B_x C/SiC$ followed by RT42A and an external glaze. Those specimens with a conversion coating (applied by HITCO) were conversion coated after the surface preparation had been performed and prior to the application of the CRT multilayer CVD coating. Surprisingly, when a conversion coating is applied to the surface-prepared substrate, the material does not perform as well as it does without a conversion coating.

COLLECTED PERFORMANCE RESULTS

To provide the reader with a good visual comparison of the separate effects discussed in this paper, all performance results have been combined on a single plot in Figure 17. As has been demonstrated in other oxidative environments[6], specimen life increases with thicker coatings. From this figure the relative effects of coating chemistry and surface preparation are apparent.

CONCLUDING REMARKS

A number of factors influence the performance of carbon-carbon oxidation protection systems. Most notably, certain variations in coating chemistry, particularly boron, appear to improve the performance of the OPS, whereas other variations, such as changes in the silicon-based coating layers, i.e. silicon-rich SiC or amorphous Si_3N_4, appear to have little influence on performance. Within a coating system of specified chemistry, lifetime is increased by increased coating thickness. No differences in performance were observed for coatings on substrate faces or edge surfaces of 2-D substrates. Nor were any appreciable differences observed for 2-D or 3-D angle interlock substrate materials.

A significant effect was observed, however, from specimen surface preparation. The preparation technique employed entails, at least in part, a geometrical modification of the surface. An initial evaluation of the surface stress state induced by geometric surface modifications indicate that these changes can account for the improved performance of these materials in the oxidizing enviroment.

REFERENCES

1. Curry, D. M., "Carbon-Carbon Materials Development and Flight Certification Experience from Space Shuttle," Proceedings of a Workshop held at Langley Research Center, Hampton, Virginia, September 15-16, 1987, NASA Conference Publication 2501.
2. Webb, R. D., 1985, "Oxidation-Resistant Carbon-Carbon Materials," Proceedings of a Joint NASA/DOD Conference, Cocoa Beach, Florida, January 23-25, 1985, NASA Conference Publication 2406.
3. Jawed, I., and Nagle, D. C., 1986, "Oxidation Protection in Carbon-Carbon Composites," Mat. Res. Bull., Vol. 21, pp.1391-1395.
4. Cullinan, J., Schaeffer, J., Gulbransen, E. A., Meier, G. H., and Petit, F. S., "Second Annual Report on Program to Study the Oxidation of Carbon-Carbon Composites and Coatings on these Materials", AFOSR Grant No. AFOSR-86-0251, August 31, 1988.
5. Kidd, Richard W. (Science Applications International Corporation, Pacoima, California) personal communication.
6. Sheehan, J. E., 1987, "Oxidation-Protected Carbon-Carbon Composites Development at GA Technologies, Inc.," Proceedings of a Workshop held at NASA Langley Research Center, Hampton, Virgina, September 15-16, 1987, NASA Conference Publication 2501.
7. Engineering Property Data on Selected Ceramics, Volume II. Carbides, MCIC-HB-07-Vol II, 1979, Metals and Ceramics Information Center, Battelle Columbus Laboratories, Columbus, Ohio, 1979.
8. Engineering Property Data on Selected Ceramics, Volume I, Nitrides, MCIC-HB-07-Vol. I, 1976, Metals and Ceramics Information Center, Battelle Columbus Laboratories, Columbus, Ohio, 1979.
9. Scott, R. O., Shuford, D. M., Webster, C. N., and Payne, C. W., 1982, NASA Contractor Report 165842-1.
10. Ohlhorst, C. W. and Ransone, P. O., 1987, NASA Technical Paper 2734.
11. NASA Langley, unpublished work.
12. Maahs, H. G., Ohlhorst, C. W., Barrett, D. M., Ransone, P.

O., and Sawyer, J. W., 1988, "Response of Carbon-Carbon Composites to Challenging Environments", <u>Materials Research Society Symposia Proceedings</u>, Vol. 125, Reno, NV, April 5-9, 1988, Materials Research Society.

13. Copley, D. C., and Rooney, M., 1987,"Non Destructive Evaluation of Carbon-Carbon Coatings," AFWAL-TR-87-4086, Oct., 1987.

14. NASA Langley, unpublished work.

15. Dietrich, H. (Chromalloy Research and Technology Division, Orangeburg, New York), personal communication.

16. Stroud, C. W. and Rummler, D. R., 1981, "Mass Loss of TEOS-Coated RCC Subject to the Environment of the Shuttle Wing Leading Edge," NASA Technical Memorandum 83203.

17. Williams, R. M., 1986, "National Aero-Space Plane: Technology for America's Future," <u>Aerospace America</u>, Vol. 24, p. 18.

18. Price, R. J., Gray, P. E., Engle, G. B. and Sheehan, J. E., 1987, "Advanced Oxidation-Inhibited Matrix Systems for Structural Carbon-Carbon Composites," AFWAL-TR-87-4047, June, 1987.

19. Singerman, S. A., Warburton, R. E., 1988, "Environmental Effects on Coated Inhibited Carbon/Carbon Composites," <u>Proceedings of the Workshop held at NASA Langley Research Center</u>, Hampton, VA, September 15-16, 1987, NASA CP-2501.

20. Sheehan, J. E., "Ceramic Coatings for Carbon Materials," GA Technologies Report GA-A18807, April, 1987.

21. Kamego, A. A. (LTV Missiles and Electronics Group, Dallas, Texas), personal communication.

22. Evans, A. G., Drory, M. D., and Hu, M. S., 1988, "The Cracking and Decohesion of Thin Films," <u>J. Mater. Res.</u>, Vol. 3, pp. 1043-1049.

23. NASTRAN Users Manual, Version 654, Document No. MSR-39, MacNeal-Schwendler Corporation, November, 1985.

TABLE 1. DESCRIPTION OF SPECIMENS

SPECIMEN NUMBER	SUBSTRATE				COATING		EXTERNAL GLAZE	SURFACE PREPARATION
	SOURCE	TYPE	INHIBITOR	STRUCTURE	PROCESS	COMPOSITION		
1	LTV[a]	ACC	No	2D	Type III	SiC	None	None
2	LTV	ACC	No	2D	Type IV	B + SiC	None	None
3	LTV	ACC	No	2D	Type VI	B + SiC	None	None
4	LTV	ACC	No	2D	Type IV plus CRT	B + SiC	None	None
5	LTV	ACC	No	2D	Type IV plus MID[d]	B + SiC	None	None
6	LTV	ACC	No	2D	Type VI plus CRT	B + SiC	None	None
7	SAIC[b]	ICCS	Yes	2D	MOD IV	B_4C + SiC	None	None
8	SAIC	ICCS	Yes	2D	MOD IV	B_4C + SiC	None	None
9	SAIC	ICCS	Yes	2D	MOD IV	B_4C + SiC	None	None
10	SAIC	ICCS	Yes	2D	MOD IV	B_4C + Si_3N_4	None	None
11	SAIC	ICCS	Yes	2D	MOD IV	B_4C + Si_3N_4	None	None
12	CRT[c]	Rohr	Yes	2D	CVD	B_xC + Si_yC	None	None
13	CRT	Rohr	Yes	2D	CVD	B_xC + Si_yC	Yes	None
14	CRT	Rohr	Yes	3D	CVD	B_xC + Si_yC	None	None
15	CRT	Rohr	Yes	3D	CVD	B_xC + Si_yC	Yes	None
16	CRT	Rohr	Yes	3D	CVD	B_xC + Si_yC	Yes	None
17	CRT	CC137E	Yes	2D	CVD plus Conversion	B_xC + Si_yC	Yes	Yes
18	CRT	CC137E	Yes	2D	CVD plus Conversion	B_xC + Si_yC	Yes	Yes
19	CRT	CC137E	Yes	2D	CVD	B_xC + Si_yC	Yes	Yes
20	CRT	CC137E	Yes	2D	CVD	B_xC + Si_yC	Yes	Yes

[a] LTV = LTV Aerospace and Defense Company
[b] SAIC = Science Applications International Corporation
[c] CRT = Chromalloy Research and Technology Division
[d] MID = Midland Materials

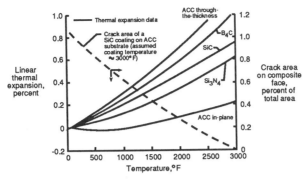

Figure 1. Linear thermal expansion of ceramics used in oxidatively protected carbon-carbon and calculated crack area of a SiC coating on an ACC substrate as functions of composite operating temperature.

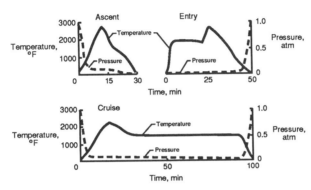

Figure 2. Representative mission simulation test environments.

Capabilities

- 3 independent test stands
- Computer control
- 18 specimen capability
- Temperatures to 3000°F
- Pressures 1.3×10^{-8} to 1 atm
- Load to 2000 lbs
- Adjustable atmospheric composition

Figure 3. NASA Langley Research Center Multiparameter Environmental Simulators.

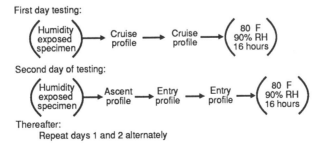

Figure 4. Specimen oxidative test plan which includes humidity exposure.

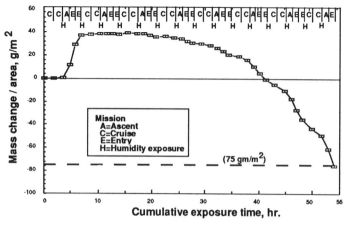

Figure 5. Typical oxidative testing results for oxidation resistant carbon-carbon composite specimen.

Figure 6. Photomicrograph of a coated carbon-carbon specimen showing variations in coating thickness.

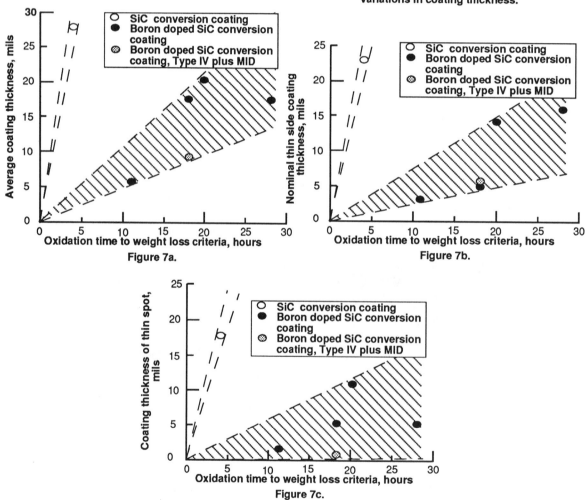

Figure 7a.

Figure 7b.

Figure 7c.

Figure 7. Plots of coating thickness versus oxidation time to weight loss criteria (75 g/m^2) for LTV uninhibited ACC with no externally applied sealants or glazes. a. Average coating thickness, b. Nominal thin side coating thickness, and c. Thickness of thinnest measured spot.

294

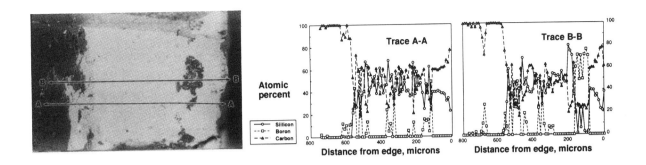

Figure 8. Elemental analyses through a Type IV coating showing the boron, carbon, and silicon molar concentrations.

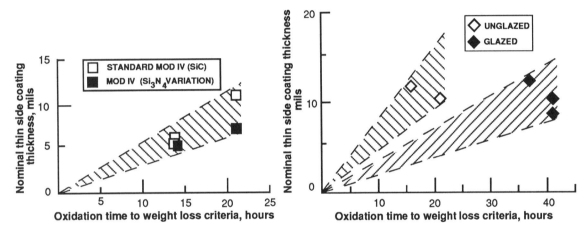

Figure 9. Comparison of SiC versus Si_3N_4 substituted MOD IV coatings on SAIC ICCS 2-D substrates.

Figure 10. Performance of glazed and unglazed carbon-carbon.

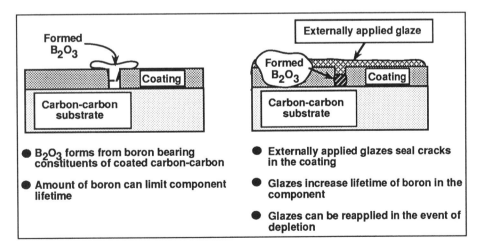

* B_2O_3 forms from boron bearing constituents of coated carbon-carbon

* Amount of boron can limit component lifetime

* Externally applied glazes seal cracks in the coating

* Glazes increase lifetime of boron in the component

* Glazes can be reapplied in the event of depletion

Figure 11. Schematic of glazed and unglazed oxidation protection systems.

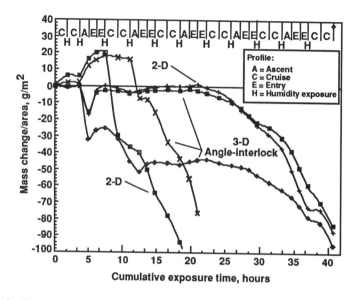

Figure 12. Performance comparison of 2D versus 3D substrate materials.

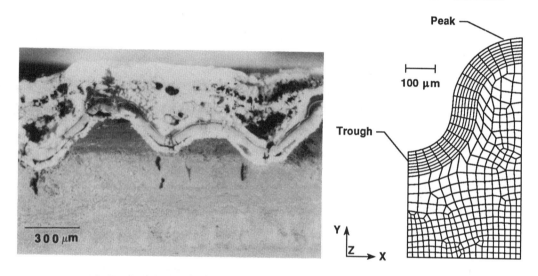

Figure 13. SEM photomicrograph of cross-sectioned, post-oxidized CRT surface prepared specimen (left side) and grid used to perform two dimensional finite element analysis of the stresses in this coating.

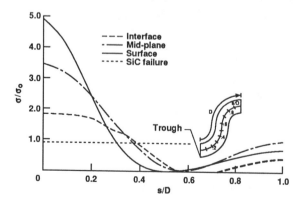

Figure 14. In-plane stresses produced in the coating.

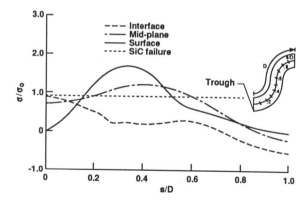

Figure 15. Normal stresses produced in the coating.

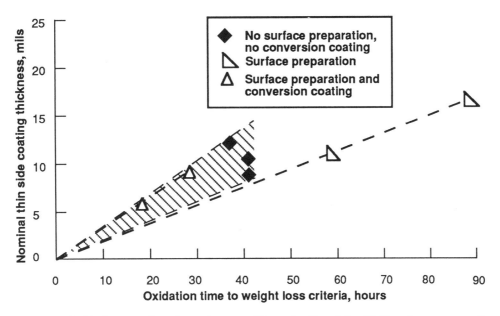

Figure 16. Oxidation results of specimens wiith and without the CRT surface preparation

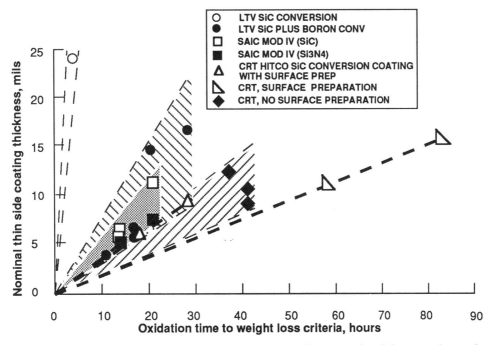

Figure 17. Summary of tested specimens showing the influence of variables on the performance of coated carbon-carbon composites in a hot aerostructural environment.

Author Index